HARPER
DICTIONARY
OF
ELECTRONICS

Ian R. Sinclair

Series Editor, Eugene Ehrlich

HarperPerennial
A Division of HarperCollins*Publishers*

This book is dedicated to Mrs. Eileen Murphy, whose
husband, Daniel Murphy, was my
colleague and friend for more than 30 years.
Dan's death came shortly after completion
of this, our final collaboration.

Eugene Ehrlich

ISBN 0-06-271528-3
ISBN 0-06-461022-5 (pbk.)
Library of Congress 90-56001

91 92 93 94 95 CC/AG 5 4 3 2 1

PREFACE

Although the foundations of electronics were laid in the late nine-
teenth century, the technology that we know as electronics is very
much a product of the twentieth century, and the most rapid
development of electronics has occurred in the years following the
1950s. Prior to that time, the word electronics had not been
coined and the topics that we now consider as being part of electronics were then classed as part of radio engineering. Electronics
has changed from being a development of radio engineering into
an all-pervading subject in its own right, spawning various subdivi-
sions and tangential topics as its growth exceeded that of any other
engineering science. The rapid rise in the use of electronic devices
has been brought about by the ubiquitous silicon crystal and the
integrated circuit, products of the years since World War II. The
integrated circuit itself was a by-product of the space race, one
whose effect has been much further-reaching than the landing of
a man on the moon.

As happens with any science that has been the subject of a
rapid expansion, explanations and definitions have lagged behind
the development of the technology. Each new development has
brought with it a huge quantity of new terms, some of them slang
words that have vanished almost as soon as they appeared, others
of lasting value, and a few that, like *feedback*, have been assimilat-
ed into everyday English, though usually in an incorrect sense.
Although the rate of growth of electronics has not really abated,
the flow of new terms has been reduced to a level at which it
becomes easier to take stock and select which terms are of lasting
use and which were transient. There still are dictionaries that present slang words of the World War II period as if they were a part
of modern electronics; in contrast, the present volume is an
attempt to define and explain the terms I believe to be relevant to
modern electronics.

The book is intended to assist anyone who needs to know the
definitions of electronics terms. This, of course, includes the student, whether at school or in the early years of a university or col-

PREFACE

lege course, but it is important to remember the needs of the non-specialist. To an ever-increasing extent, electronic devices and methods are impinging on studies that at one time would have appeared to be unlikely users of such devices. From archaeology to music, from medicine to navigation, electronic measurements and methods now need to be understood by students and practitioners of subjects that appear to have very little in common. The need for, and use of, electronics now transcends all the artificial divisions of subject matter, and makes a dictionary of electronics that is expressed in plain language a necessity rather than a luxury.

This book has therefore been designed with more than the needs of the traditional electronics student in mind. Each explanation is couched in accessible terms, avoiding the use of mathematics, except where a note on the mathematical aspect of a topic is necessary as a reminder to the prospective designer as distinct from the user of electronics. The form of the dictionary has been set out so as to make it easier to find an explanation of a device or principle under one heading rather than requiring the user to flit from one heading to another, picking up fragments of learning from each. Additionally, cross-references (indicated by SMALL CAPITAL LETTERS) guide the user to related entries. I am most grateful to Ian Crofton and Edwin Moore, of HarperCollins*Publishers*, for their unstinting efforts in working this book into its final form from the material on floppy disks, for it is their efforts that have made this a readable and usable dictionary. I am particularly grateful to Dr. John Turner of Southampton University for reviewing the manuscript, and for his many helpful suggestions, elucidations, and revisions.

Ian Sinclair

A

aberration any distortion of an image by a lens system. In electronics the term denotes distortions in the electron lenses of TV CAMERA TUBES, CATHODE-RAY TUBES, ELECTRON MICROSCOPES, etc. See also BARREL DISTORTION, PINCUSHION DISTORTION, ELECTROSTATIC LENS, ELECTROMAGNETIC LENS.

abrupt junction a JUNCTION (2) with a sudden change of MAJORITY CARRIER, in which one material is lightly doped, either as DONOR or ACCEPTOR, with the other material heavily doped in the opposite polarity.

absolute scale or **Kelvin scale** the scale of temperature that uses ABSOLUTE ZERO as its zero point.

absolute temperature a temperature measured in the ABSOLUTE SCALE, with ABSOLUTE ZERO at about −273°C. All fundamental equations in electronics that make use of temperature will require the units of temperature to be in the absolute scale.

absolute value circuit a circuit that gives a positive output equal to the magnitude of the input signal. See also BRIDGE RECTIFIER.

absolute zero the zero of the ABSOLUTE SCALE, equivalent to about −273°C. The temperature is calculated theoretically from the behavior of gases and is the point at which THERMAL NOISE ceases. In practice, the value is confirmed from the fact that the absolute zero temperature can never be attained exactly, though it is possible to reduce temperatures to within a fraction of a degree of absolute zero.

absorption the extraction of energy, usually from a WAVE. A radio CARRIER WAVE is absorbed by many materials, resulting in the conversion of the absorbed energy into heat or some other form of energy. The alternatives to absorption are TRANSMISSION, reflection and refraction.

absorption loss a measure of the amount of power loss caused by

ABSORPTION. The absorption loss is usually quoted in DECIBELS
and is equal to 20 log(power lost/power input).

AC see ALTERNATING CURRENT.

AC bias an electrical signal used in analog audio tape-recording
systems. A graph of the retained magnetism (REMANENCE) of a
magnetic material plotted against the AMPERE-TURNS of the magnetizing coils is never a straight line. When a high-frequency sine
wave AC signal is applied to the recording head, the lower frequency signal to be recorded can be mixed with this signal. The
result is that the signal to be recorded causes values of remanent
magnetism to be generated that lie on a comparatively straight
part of the graph. On playback, the high-frequency signals can be
filtered out.

The use of digital recording methods is a better solution. See
RECORDING OF SOUND.

AC generator any device or circuit that generates ALTERNATING
CURRENT. The term usually denotes a rotating machine, such as
an ALTERNATOR. Purely electronic circuits that generate AC are
more commonly described as waveform generators or SIGNAL
GENERATORS.

accelerated life test a LIFE TEST conducted under unusually
harsh conditions. This normally involves use of a supply voltage
higher than normal, greater than normal power dissipation, or
repeated operation of switching devices. The goal is to determine
whether samples have consistent lifetimes, not to predict actual
life under normal conditions. If sufficient statistical evidence is
available, however, it may be possible to correlate expected life
with the results of accelerated life tests.

accelerating anode a metal cylinder that is maintained at a positive voltage relative to the CATHODE of a CATHODE-RAY TUBE.
The resulting electric field accelerates the electrons to high
speeds, but because of the shape of the anode, few electrons strike
it. The current that flows to the electrode is therefore negligible
compared with the current to the FINAL ANODE, which is normally
the display screen.

acceleration voltage the steady voltage, measured with respect
to the CATHODE, that is applied to the accelerating anode.

acceptance angle the maximum angle, measured relative to the optical axis of a PHOTOCELL, at which light can strike the cell and still cause an output.

acceptor a P-TYPE material for DOPING a pure semiconductor that accepts electrons into its structure, thus releasing positive charges (HOLES) in the semiconductor. Compare DONOR.

access time a delay in MEMORY response. When a memory is addressed by placing address bits onto the ADDRESS BUS little time will elapse before the memory makes connection with the data bus for reading or writing. This time is the access time. For modern memory chips in microcomputers, access times of the order of 200ns or less are usual. Much shorter access times are required for larger machines using magnetic CORE STORES.

accumulator 1. the main REGISTER of a microprocessor. Each of the operations of addition, subtraction, AND, OR, and XOR will affect a byte in the accumulator and a byte taken from memory or a secondary register. The result of the action is then stored in the accumulator. 2. a form of secondary storage cell (see LEAD-ACID CELL, SECONDARY CELL).

AC/DC 1. *abbreviation for* alternating current/direct current. 2. *adj.* (of electric motors and electronic equipment) able to operate on either alternating current or direct current supply. The term was originally applied to radios and TV receivers that could be connected directly to AC power lines and used no transformers. Nowadays, the term is usually applied to equipment that can be operated either from AC power lines, using a transformer, or from low-voltage DC, such as a car battery.

AC motor any form of motor that requires an AC supply, that is, all types of INDUCTION MOTORS, SYNCHRONOUS MOTORS, and SHADED-POLE MOTORS. Compare UNIVERSAL MOTOR.

acoustic delay line a method of delaying electrical signals by converting them into ULTRASONIC signals. A TRANSDUCER converts the electrical signals into acoustic, or vibratory, signals. These travel through a material at the speed of sound, which is much slower than the speed of electromagnetic waves, and are reconverted into electrical signals by another transducer. The original acoustic delay lines used mercury as the wave medium. Modern acoustic delay lines are manufactured from glass for use in COLOR TV receivers to equalize the time taken by color signals, which have different bandwidths. Acoustic waves are also used in wave FILTERS. See DELAY LINE, SURFACE ACOUSTIC WAVE.

acoustic feedback the unwanted feedback of sound waves. The most common form is the squealing known as HOWL, which is caused when a public address system is used with a loudspeaker placed too close to a microphone. This is a form of POSITIVE FEEDBACK in which the soundwave from the loudspeaker reaches the microphone to be amplified over again, resulting in oscillation. The cure is to use strongly directional microphones, or circuits ensuring that the phase of acoustically fed-back signals is always incorrect for oscillation. Acoustic feedback can also affect phonograph PICK-UPS placed too close to loudspeakers.

acoustic wave any compression wave in a material in the form of a sound wave, whether or not the result is audible.

acoustic wave device any electronic device that makes use of acoustic waves. See SURFACE ACOUSTIC WAVE, ACOUSTIC DELAY LINE.

AC power analyzer a measuring instrument that detects and records deviations from the normal sine wave power supply.

AC resistance the complex resistance of an AC circuit.

In an AC circuit there may be a phase shift between voltage and current. For this reason the resistance of an AC circuit cannot be specified, since both the magnitude and PHASE of voltage and current are required. A COMPLEX NUMBER form of OHM'S LAW can be defined as $V = IZ$ where V is complex voltage, I is complex current, and Z is complex IMPEDANCE. AC resistance R is then the real part of impedance, which can be expressed as

$$Z = R + jX,$$

where X is REACTANCE.

activated cathode a hot CATHODE of the oxide-coated type. See THERMIONIC CATHODE.

active antenna a transmitting antenna that is supplied with radio-frequency power. See also ANTENNA, PARASITIC ANTENNA, DIRECTOR.

active area the area in a rectifier JUNCTION that actually passes current.

active component a circuit component capable of increasing the power of a signal. Also called *active device.* An active component requires a power supply in order to operate and can have a measurable POWER GAIN as distinct from voltage gain or current gain. Compare PASSIVE COMPONENT.

active current see IN-PHASE COMPONENT.

active device see ACTIVE COMPONENT.

active filter a form of FILTER that uses ACTIVE COMPONENTS, usually OPERATIONAL AMPLIFIERS. The PASSIVE COMPONENTS that determine the filter's performance are part of a FEEDBACK CONTROL LOOP around the active device. See Fig. 1.

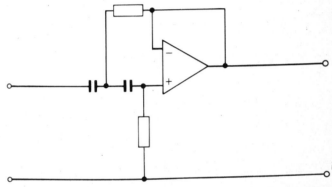

FIG. 1. **Active filter.** A typical active filter circuit, in this example a high-pass filter. (The power-supply connections to the operational amplifier are not shown.)

active load an active device used as a power dissipating load. Since a TRANSISTOR whether bipolar or FET, is a form of variable resistor, it can be used as a load whose resistance value is variable.

active network a set of components that includes an ACTIVE COMPONENT. The term usually denotes any FILTER network that includes a transistor, operational amplifier, or other device with power gain.

active transducer a form of transducer with power gain. Most transducers are PASSIVE COMPONENTS but a transducer based on a transistor, such as a PHOTOTRANSISTOR, can supply power gain in addition to energy conversion.

active voltage see IN-PHASE COMPONENT.

ACTIVITY

activity a measure of the efficiency of a QUARTZ CRYSTAL. A crystal oscillator provides a driving voltage that sets the crystal into natural oscillation, or resonance. The peak signal voltage across the crystal is many times greater than the driving voltage, and the ratio of peak crystal voltage to peak driving voltage is the activity figure of the crystal under these conditions. Activity figures in the tens of thousands are common.

actuator any electrically operated device that performs a mechanical action. The term is used extensively in control systems and ROBOTICS. An actuator can be a simple solenoid, or an elaborate system of motors.

adder a LOGIC CIRCUIT that provides the binary adding action summarized by the truth table in Fig. 2. An adder circuit that provides for a carry bit *into* the addition is called a FULL ADDER and if this carry-in is not provided for, the device is called a HALF-ADDER.

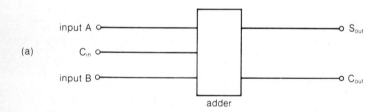

(a)

adder

(b)

A	B	C_{in}	S	C_{out}
0	0	0	0	0
0	0	1	1	0
0	1	0	1	0
0	1	1	0	1
1	0	0	1	0
1	0	1	0	1
1	1	0	0	1
1	1	1	1	1

FIG. 2. **Adder.** (a) Circuit symbol. (b) Truth table for a two-bit full adder.

address a number used to activate a MEMORY LOCATION or a PORT connected to a MICROPROCESSOR. The microprocessor controls other parts of the system by using a set of connections called the ADDRESS BUS. Each line in an address bus links the microprocessor to each of the memory and port units in a system. When a set of binary signals is placed on the address bus by the microprocessor, the bus is used to locate one unique port or part of memory. Each address number therefore corresponds to one memory unit or one port. When a device is addressed by putting its address number on the bus, it can then be read from—data copied *from* the device to the microprocessor—or it can be written to—data copied from the microprocessor *to* the device.

address bus the set of connections that carries binary ADDRESS numbers to all parts of a microprocessor system.

admittance the reciprocal of IMPEDANCE. Symbol: **Y**; unit: SIEMENS (formerly *mho*). For circuits in which it is more convenient to work with currents than with voltages, the equation $Y = G + jB$ is used rather than the more familiar $Z = R + jX$. The symbol G is used for conductance (1/resistance), and the symbol B is used for susceptance (1/reactance). See COMPLEX NUMBER.

admittance gap the hole in a WAVEGUIDE or CAVITY RESONATOR at which an electron beam and a resonant microwave signal can interact.

AF see AUDIO FREQUENCY.

AFC see AUTOMATIC FREQUENCY CONTROL.

afterglow the light pattern visible from the PHOSPHOR of a CATHODE-RAY TUBE screen after the electron beam has been cut off. For radar use, a long afterglow, of several seconds, is useful in allowing the tracks of moving objects to show up as a trail. For TV use, the afterglow must be short enough to avoid any appearance of trail. At the same time, some afterglow is desirable to avoid excessive FLICKER.

AGC see AUTOMATIC GAIN CONTROL.

aging or **burn-in** the operating of a new component or circuit until its characteristics become stable. The term is usually applied to transmitting VACUUM TUBES and other HOT-CATHODE vacuum devices, in which small changes occur in the device characteristics during the first few hundred hours of life. This process can sometimes be speeded up by using severe conditions, such as raised voltage supply, higher input power, and higher operating temperature. Some types of PASSIVE COMPONENTs can also benefit from aging.

air-break switch a switch whose contacts are designed to separate in air, rather than in a vacuum or in oil.

air capacitor a capacitor that uses air as its only DIELECTRIC. The device is used for some variable capacitors and for some high-voltage transmitter capacitors. See also VACUUM CAPACITOR.

air gap any air-filled discontinuity in a material, particularly in MAGNETIC CORES.

Alcomax *Trademark.* a magnetic material consisting of an alloy of iron, nickel, aluminum, cobalt, and copper that has exceptionally high coercive force (see MAGNETIC HYSTERESIS). It is used mainly in making permanent magnets for loudspeakers and for MAGNE-TRONS.

aliasing an effect produced when a signal is sampled too slowly, leading to a false result. When a varying analog signal is measured or sampled at regular intervals, for example, by an analog-to-digital converter, SHANNON's SAMPLING THEOREM states that the sample rate must be more than twice as fast as the rate of signal variation if aliasing is to be avoided. An illustration of the false results produced by aliasing can be seen in films of stagecoaches and wagons, in which the wheels appear to revolve slowly backward. This is because the motion of the spokes was sampled too slowly by the motion picture camera.

alloyed junction a JUNCTION (2) produced by depositing metal on a piece of semiconductor material, which is then heated to form an alloy. The method was once used to produce germanium transistors of better high-frequency characteristics than the normal diffused junction type.

all-pass network a circuit that attenuates signals of all frequencies equally. Compare FILTER.

alnico a magnetic material, an alloy of aluminum, nickel, and iron with copper and cobalt that is used for permanent magnets.

alpha cutoff the conventional limit of operating frequency for a transistor. COMMON-BASE CONNECTION of a transistor enables operation up to the highest possible frequency the construction of the transistor permits. The alpha cutoff frequency is the frequency at which the voltage gain of the transistor in a common-base circuit has fallen to 0.707 of the voltage gain at lower frequencies.

alphanumeric (of a character set, code, or file of data) consisting of alphabetical and numeric symbols.

alternating current (AC) a current that flows in each of two directions alternately. The term usually implies that the time between current reversals is constant, so the FREQUENCY of the current reversal is constant. The WAVEFORM of current plotted against time is also assumed to be symmetrical around zero and is usually a sinusoid. Alternating current is normally used to denote the *line power supply.* See also PULSE, MARK-SPACE RATIO.

alternator a mechanical generator of ALTERNATING CURRENT. An alternator may be thought of as a coil of wire rotating in a constant magnetic field. If the connections to the ends of the coil are made through slip rings to ensure continuous contact, the waveform that is supplied will be *alternating current.*

For example, if a single-turn square coil with sides of length a is rotated at angular velocity ω about an axis perpendicular to a uniform magnetic field $\mathbf{B}$ then the flux through the coil at time t will be

$$\phi = \mathbf{B}a \, \text{Sin} \omega t$$

and the voltage generated at the terminals of the coil will be

$$V = \frac{-\mathrm{d}\phi}{\mathrm{d}t} = - \omega \mathbf{B} a \, \text{Cos} \omega t.$$

aluminized screen a CATHODE-RAY TUBE screen whose brightness is increased by coating the PHOSPHOR with a thin film of aluminum, slightly spaced from the phosphor. Light from an excited part of the phosphor that would normally shine back inside the tube is thus reflected forward, increasing the total light output from the screen. The aluminum film must be thick enough to reflect efficiently, but not so thick as to absorb too great a proportion of the electrons that strike it.

AM see AMPLITUDE MODULATION.

amateur bands the SHORTWAVE bands in the radio spectrum reserved for use by licensed amateur operators. At one time, the shortwave bands were regarded as useless, and it was only the experiments of these radio hams that demonstrated that the bands in fact were the most efficient method of long-distance transmission available before the use of satellites. Governments then took possession of the shortwave bands, but selected parts were made available to amateurs in recognition of their achievements. Licensed amateurs continue to make significant contributions to the development of radio communication.

AM/FM radio a radio that can receive both AMPLITUDE MODULATED and FREQUENCY MODULATED signals.

ammeter the ampere meter, a meter used for measurements of current. For most purposes, ammeters use MOVING-COIL METER movements, shunted (see SHUNT) to the appropriate range. AC ammeters generally employ RECTIFIERS along with moving-coil movements. THERMOAMMETERS, hot-wire ammeters, are used for measuring radio frequency currents, such as antenna currents.

amp see AMPERE.

ampere (amp) the unit of electric current, defined in terms of the force between a pair of wires. Symbol: **A**.

The definition implies two infinitely long, straight, parallel wires of negligible diameter, suspended one meter apart in a vacuum, and with an identical current flowing in each wire. When the force between the wires is 2×10^{-7} newtons per meter length of wire, then the current flowing in the wires is one ampere. This definition is another way of stating that the BIOT-SAVART LAW is held to be the defining equation for magnetic FLUX DENSITY.

The ampere is named for André Marie Ampère (1775—1836), the French physicist and mathematician who made major discoveries in the fields of magnetism and electricity.

ampere per meter the unit of MAGNETIC FIELD strength. Symbol: H.

Ampère's law see BIOT-SAVART LAW.

ampere-turn the unit of MAGNETOMOTIVE FORCE (mmf) for a coil. The mmf of a coil is found by multiplying the number of turns by the current flowing. The size of magnetomotive force can then be related to the flux density by using the equation mmf = reluctance × flux.

amplification see GAIN.

amplification factor the absolute value of small signal GAIN. The amplification factor for a vacuum tube or other device is defined as change of output voltage divided by change of input voltage, disregarding sign. The quantity has not been as widely used for semiconductor devices.

amplified Zener a form of VOLTAGE STABILIZER circuit. The circuit uses a ZENER DIODE with a power transistor connected to carry the main stabilizing current. See Fig. 3.

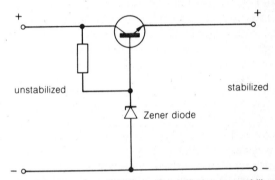

FIG. 3. **Amplified zener.** One form of amplified zener stabilizer, in which the transistor carries the stabilized current. The zener diode provides the stable base voltage for the transistor. Another form of amplified zener circuit is shown with the entry for SHUNT STABILIZER.

amplifier a circuit that provides power GAIN. Amplifiers can be classified as voltage amplifiers, current amplifiers, transconductance amplifiers, or transresistance amplifiers according to their design and use.

A voltage amplifier amplifies voltage while providing a constant current. A current amplifier amplifies current at a constant voltage. A transconductance or transresistance amplifier is one in which the output current (or voltage) has a well-defined relation to the input voltage (or current).

amplifier stage a single section of an amplifier consisting of several sections.

amplitude the size of a wave at a given time. For a voltage wave, this will mean the voltage of the wave, whether positive or negative, at an instant. The peak amplitude is the amplitude measured from zero to one peak of the wave. The peak-to-peak amplitude, as measured by an oscilloscope, is the amplitude measured between the positive peak and the negative peak of the wave.

The term 'amplitude' is often reserved for the quantity A in

$$x = A \operatorname{Sin} (\omega t + \phi),$$

where x is the instantaneous amplitude at time t, A the peak amplitude, ω the ANGULAR FREQUENCY and ϕo the PHASE ANGLE.

amplitude modulation (AM) a form of encoding that provides a means of carrying information by a radio wave. The AMPLITUDE of a high-frequency (radio) wave (see CARRIER 1) is increased or decreased by the amount of the amplitude of a low-frequency wave (audio or video). This results in the outline, or ENVELOPE of the high-frequency wave taking the shape of the low-frequency wave. See also FREQUENCY MODULATION, SINGLE SIDEBAND, VESTIGIAL SIDEBAND.

analog circuit a circuit whose output signal voltage is proportional to its input signal voltage, That is, any circuit, supplied with a varying input of suitable frequency, will produce an output whose voltage at any time bears a relationship to the voltage at the input.

analog computer a computer in which analog voltages are used to represent quantities, and active circuitry is used to represent mathematical operations. The analog computer uses the AMPLITUDE of a wave to represent the size of a quantity, making linearity of amplification extremely important (see LINEAR AMPLIFIER). Mathematical actions such as integration and differentiation are represented by the correspondingly named electrical operations on a waveform. The basic circuit of an analog computer is the OPERATIONAL AMPLIFIER a very high-gain linear amplifier whose characteristics can be modified by using NEGATIVE FEEDBACK. Compare DIGITAL COMPUTER.

analog/digital converter (ADC) a circuit that converts the analog voltages or currents into digital, usually binary, numbers. The digital signal gives the amplitude of the analog signal at a particular instant.

A common way of achieving an A-to-D conversion is to measure the time taken for a steadily increasing voltage to equal the input voltage. This time is then converted into binary code. The rate at which conversions are carried out must be considerably greater than the frequency of the wave being sampled. See also QUANTIZATION, ALIASING.

AND a logical operation in which a TRUE output is given only if two or more inputs are TRUE. See OPEN-COLLECTOR DEVICE.

AND gate an electronic circuit whose action carries out AND logic. The output of the AND gate is TRUE (logic level 1) only when each input is simultaneously TRUE. See Fig. 4.

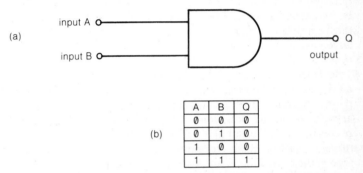

FIG. 4. **AND gate.** (a) Circuit symbol. (b) Truth table for the AND gate. If more than two inputs exist, the logic is that the output is 1 only if all inputs are 1.

anemometer a gauge of wind strength. Electronic anemometers have largely replaced the older types for navigational purposes. The simplest system uses a TACHOGENERATOR driven by a propeller to measure wind speed, and a wind vane mounted on a potentiometer to transmit directional signals to the receiver. The advantage of electronic methods is that the distance between transmitter and receiver is unimportant.

angle of flow the fraction of a radio-frequency CYCLE for which current flows in an amplifying device. The term is used mainly for CLASS C transmitting stages. Conduction at all parts of the cycle corresponds to an angle of flow of 360 degrees, conduction for half of a cycle to an angle of flow of 180 degrees, and so on. Class C amplifiers normally have angles of conduction of considerably less than 180 degrees.

angular frequency a wave frequency expressed in radians per second. One cycle of a wave is considered as the projection onto a line of one revolution of a circle, with one revolution tracing out the angle 2π. The angular frequency is therefore the frequency in hertz multiplied by 2π.

angular modulation a form of modulation in which the PHASE ANGLE of the CARRIER is affected by the amplitude of the modulating signal. Angular modulation is often used as a means of frequency modulation, and an angular modulated signal can usually be demodulated by an FM radio receiver.

anion a negatively charged ION. The term is so called because such an ion will move, if free to do so, to a positively charged electrode or ANODE. Compare CATION.

anisotropic (of a material) having CHARACTERISTICS that depend on direction. A simple example is wood, in that its strength depends on grain direction. In electronics, the term is applied to crystals and to magnetic materials whose magnetic or electrical properties are different along different axes.

anode or **plate** a conductor held at a positive potential. In electronics the term denotes an electron collector in a vacuum device, such as a vacuum tube or cathode-ray tube.

anode current the amount of electron current flowing from the electron source (CATHODE) to the anode. Since the conventional direction of current flow is from positive to negative, this is described as a current flowing from anode to cathode.

anodizing the protecting of aluminum with a skin of oxide, which also serves as an electrical insulator but not as a good heat insulator. Anodized aluminum surfaces can therefore be used as HEAT SINKS that also provide electrical insulation.

ANSI a US organization that recommends standards for many products in various industries.

antenna a transmitting or receiving device for radiated waves. The antenna acts as a type of MATCHING transformer for waves along a line and waves in space so that the maximum transfer of energy can be achieved. Achieving high efficiency in an antenna system is of great importance both for transmitting and for receiving.

Because of the great differences in the wavelengths of radio waves at the extremes of the spectrum, antennas can take a variety of shapes and sizes. For low radio frequencies, antennas take the form of long wires, well clear of the ground. For the VHF and UHF range, rod antennas cut to tuned lengths are used, and these may be of the YAGI type in order to be directional. At microwave frequencies, antennas consist of sections of launching WAVE-GUIDES with parabolic reflectors. See also DIPOLE ANTENNA, MAR-CONI ANTENNA, RADIATION PATTERN, DIRECTIVITY.

antenna array a set of antennas arranged to transmit or receive in one direction. This is done by using the correct spacing between antennas, and the correct phasing of signals, so the waves from each antenna will reinforce one another in the chosen direction.

A BROADSIDE ARRAY concentrates its energy in a direction that is at right angles to the line of antennas. An ENDFIRE ARRAY concentrates radiated energy in the direction of the antenna line. An array can be said to have a figure of gain above that of a single element, and this figure is a good measure both of efficiency and directional qualities. See also MAJOR LOBE, YAGI, RADIATION PAT-TERN, DIRECTIVITY.

antenna current the radio-frequency current in an anetnna. This is measured as the ROOT MEAN SQUARE current at the carrier frequency at some point in the ANTENNA FEEDER. For an UN-TUNED ANTENNA the current is usually measured at the feeder connection. Otherwise, the current is measured at the voltage standing wave MODE which is the position of maximum current.

antenna efficiency the percentage of power radiation efficiency of an antenna. Measured at the carrier frequency of radiation, this is the percentage of power radiatedfrom the antenna compared with power supplied to the antenna. Low antenna efficiency im-plies a poor STANDING WAVE RATIO.

antenna feeder the TRANSMISSION LINE that connects an antenna to a transmitter or to a receiver.

antenna gain a measure of the efficiency of an antenna compared with the efficiency of a standard reference antenna. The efficiency is measured in terms of the power radiated or received as com-pared with the standard under the same conditions. For VHF and UHF antennas, the standard is a simple DIPOLE ANTENNA. For lower frequencies, a theoretical infinite wire standard is sometimes used, whose radiated or received energy figures are calculated theoretically.

antenna impedance a COMPLEX NUMBER relating signal voltage to signal current at a point on the antenna. The antenna impedance will not be a constant, particularly for a TUNED ANTENNA. The impedance is therefore usually measured at the antenna FEED POINT and is often referred to as *feed-point impedance.*

antenna resistance or **radiation resistance** a quantity obtained by dividing antenna power by the square of current. If the power figure used is that of radiated power, the resulting figure is that of antenna radiation resistance. If the power figure used is that of supplied power, then the resistance figure is that of antenna resistance.

anticathode see TARGET.

anticoincidence circuit a form of XOR gate (see EXCLUSIVE OR) that has an output when there is a pulse at one input only, but no output when there are pulses at both inputs. The circuit can usually be adjusted to reject input pulses at the two inputs that are within a specified time apart. Such circuits are used in particle counters.

antihunting circuit a part of the circuitry of a SERVO control system designed to damp out oscillation around a set level. See DAMPED, HUNTING.

anti-interference antenna a receiving antenna connected to the receiver by balanced TWIN CABLE or by COAXIAL CABLE. The antenna is placed well clear of man-made interference, and the link between the antenna and the receiver is designed so that it does not pick up any signals.

antijamming any measure designed to reduce deliberate interference with electronic signals. This can include rapid changes of frequency, the use of DIGITAL coding or of highly DIRECTIONAL ANTENNAS and the use of satellite repeaters.

antinode the point in a STANDING WAVE pattern that has maximum voltage amplitude. On a power transmission line or antenna, this is the point of minimum current and maximum resistance.

antiphase a signal with a phase difference of 180°. The term is sometimes used loosely to mean *inverse* (see INVERSION), but the two are identical only for sine waves. See OPPOSITION and Fig. 5.

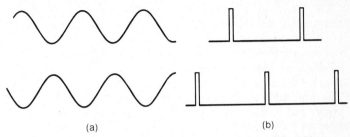

FIG. 5. **Antiphase.** For the sine wave (a) a 180° phase shift is equivalent to inversion, but for the pulse waveform (b) this is not true.

antitracking a silicone resin solution with good insulating properties. It is painted on surfaces around a high-voltage connector to avoid current leakage, or TRACKING.

antitransmit/receive (ATR) switch the antitransmit/receive switch in a RADAR waveguide. This device open circuits the transmitter waveguide during the reception period, thus steering the maximum amount of received energy into the receiver. Compare TRANSMIT/RECEIVE (TR) SWITCH.

aperiodic circuit or **nonresonant circuit** an untuned circuit, that is, any circuit that responds to a wide range of frequencies without a noticeable resonance.

aperture distortion the blurring of a TV picture due to the finite spot diameter of the electron beam. This makes it impossible to achieve a sharp transition from black to white or white to black in a TELEVISION CAMERA or PICTURE TUBE.

aperture mask or **aperture grille** a metal mask in a TV color PICTURE TUBE that is etched with a pattern of holes or narrow slits. The electron beams from the three guns of the color tube cross over at a hole or slit, and so strike the correct PHOSPHORS to achieve colored light outputs. See Fig. 6. See also COLOR CRT, SHADOW MASK.

apparent power the figure of power obtained from multiplying voltage by current. For an alternating signal, this will be equal to true power only if the phase angle is zero, which will not be the case if the circuit contains inductive or capacitive devices. *True power* is obtained when apparent power is multiplied by the cosine of the phase angle.

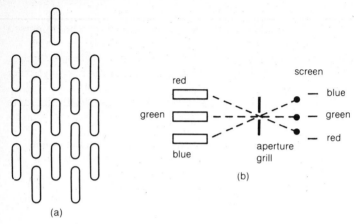

FIG. 6. **Aperture mask** or **grille.** (a) The shape of an aperture mask of a color CRT. (b) The way in which the aperture causes the beams from the three electron guns to strike the correct phosphor stripes on the face of the CATHODE-RAY TUBE.

Appleton layer see F LAYER.

array see ANTENNA ARRAY.

arc a conducting path in air. This requires the air to be ionized, and is more likely to occur with DC than with low-frequency AC because at the zero voltage portion of an AC wave, there will be no arc current, and the air can deionize in this time. An arc causes severe radio frequency interference, and evaporation of metal from the conductors between which the arc takes place.

arcing the appearance of an ARC when switch or other contacts open.

arcover the appearance of an arc between two terminals with a large potential difference. Arcover is common in all high-voltage devices, such as X-RAY TUBES, CATHODE-RAY TUBES and transmitting tubes. Arcover is prevented by insulating around high-voltage connectors, painting surfaces with ANTITRACKING material, and removing dust and moisture.

arc suppression any system for reducing ARCING. Switch contacts in DC circuits are arc-suppressed by using resistor-capacitor networks. For large switches, an airblast when the contacts open can be used to blow out the arc, or the contacts can be immersed in oil (*oil-quenched contacts*).

armature any moving piece of SOFT MAGNETIC MATERIAL such as the rotating part of an AC or DC generator or the moving part of a relay.

artificial line a network of PASSIVE COMPONENTS with the same characteristics as a TRANSMISSION LINE. See also LUMPED PARAMETER.

artwork the DIAGRAMS that will be used to form the etch patterns for PRINTED CIRCUITS or INTEGRATED CIRCUITS. These will be reproduced photographically, with any reduction of scale that is needed.

ASCII *acronym for* American Standard Code for Information Interchange, a number code used to represent letters, digits, and punctuation marks. The code is widely used in computing.

aspect ratio the ratio of width to height for a TV picture. A value of 4:3 is almost universal, but different values are usually standardized for ULTRA-HIGH DEFINITION TV systems.

assembler a computer program that can translate commands into machine code for systems software. The commands must be written in assembly language, which uses simple mnemonic abbreviations for machine code instructions.

assembly a set of components, usually on a single printed circuit board, that performs a distinct circuit action. The alternative to using assemblies is to have single-board construction.

astable circuit a circuit that can be in one of two states, neither of which is permanently stable. Astable circuits are often used as oscillators. See also MONOSTABLE, BISTABLE, MULTIVIBRATOR.

astigmatism a defect in an ELECTRON-OPTICS system in which the SPOT is elliptical rather than circular. Astigmatism in a CATHODE-RAY OSCILLOSCOPE tube makes vertical lines appear thicker than horizontal lines, or vice versa. Astigmatism in a TV CAMERA TUBE or receiver tube makes pictures appear to be out of focus in one direction.

asymmetrical waveform a waveform that has no axis of symmetry, or that has a DC component so that its axis of symmetry is a DC voltage.

asynchronous (of a signal) being unsychronized with any other signal in a circuit. Compare SYNCHRONOUS; see also UNCLOCKED.

asynchronous counter or **serial counter** a counter in which input pulses are applied only to the first stage of the counter, and the output of each stage is used as the input for the following stage. If such a counter is incremented from, say, 0111 to 1000 in binary, the stages will not change simultaneously. See also SYNCHRONOUS COUNTER.

attenuation a reduction of signal power, the opposite of GAIN. Attenuation, like gain, is measured in decibels (dB). Attenuation of signals in circuits is caused by RESISTORS and any other energy-dissipating components.

attenuation band the range of frequencies over which a FILTER attenuates signals.

attenuator a circuit that attenuates a signal by a known amount. A switched attenuator uses a resistive LADDER network and is calibrated in decibels. POTENTIOMETERS can be used as attenuators, but CALIBRATION is not as easy. See Fig. 7.

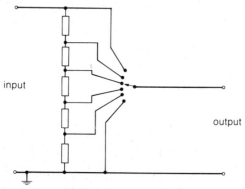

FIG. 7. **Attenuator.** A simple attenuator using a chain of resistors.

audio (of a device or frequency) making use of signals in the audio range of approximately 15Hz to 20kHz.

audio frequency (AF) the range of frequencies between about 15Hz and 20kHz. This corresponds to the range of sound wave frequencies that can be heard by a healthy human ear.

audiometer an instrument used to measure acuity of human hearing. An older type of instrument used signals into earphones and required a response from the person being tested. This is not suitable for some young children, so later models have operated on a system that uses a pulse to set the eardrum at resonance. The sound radiated by the resonating eardrum can then be picked up, recorded, and analyzed to give much more information about the ear than could be obtained by the earlier instrument.

audio signal a SIGNAL that can be used to provide a sound when applied to a LOUDSPEAKER.

autodyne an obsolete system of radio reception. The incoming carrier wave, amplified if necessary, is allowed to beat (see BEATING) with a LOCAL OSCILLATOR that is synchronized to the same frequency and phase. The BEAT NOTE extracted from this process is the original modulating frequency.

automatic control any control system that uses electrical input signals to control a system without human intervention. The electrical signals into the controller will arise from TRANSDUCERS, for example, photocells, strain gauges, limit switches, and level switches. The output will usually consist of high-power supplies for electric motors, actuators, heaters, and similar devices. Most control systems depend on application of overall NEGATIVE FEEDBACK.

automatic frequency control (AFC) a system of LOCAL OSCILLATOR control in a frequency-modulated SUPERHETERODYNE receiver. The local oscillator is made voltage dependent, usually by making a VARACTOR diode part of the tuned circuit. At the demodulator, a DISCRIMINATOR circuit gives a DC voltage that varies according to the frequency of the IF signal.

This DC voltage is connected, sometimes with a stage of amplification, to the varactor diode. If the receiver is perfectly tuned, this DC voltage should be zero. If not, then the voltage applied to the varactor diode will adjust the oscillator frequency until the receiver is in tune.

automatic gain control (AGC) or **automatic volume control (AVC)** a system for counteracting input signal amplitude fluctuations in a radio-frequency signal. The scheme applies mainly to AM signals, and makes use of the DC level at the demodulator. This DC level is proportional to the amplitude of the carrier wave and will vary as the carrier amplitude varies. By using this voltage to control the gain of several stages, usually RF and IF stages, the amount of IF signal at the detector can be made almost constant despite large variations in carrier amplitude.

automatic level control a form of AUTOMATIC GAIN CONTROL for TV signals, holding the BLACK LEVEL of the picture at a constant level.

automatic noise limiter a circuit that clips IMPULSE noise peaks from any signal. The circuit employs the principle of charging a capacitor to the peak signal voltage, and using this voltage to operate a CLIPPER. Any pulse of more than the peak amplitude is clipped.

automatic tracking a form of automatic control for a radar antenna. This keeps the transmitted beam on the target while the range is computed. Today, this is achieved by electronic means, that is, by altering the phasing of signals to different parts of the antenna, rather than entirely by control of antenna movement.

automatic tuning a form of AUTOMATIC FREQUENCY CONTROL that permits coarse setting by manual methods. If the receiver is manually put roughly into tune, the automatic system then takes over and completes the tuning process.

autotransformer a transformer with a single winding having an adjustable secondary tapping. At electrical power frequency, the device is used mainly for testing equipment that must operate with a range of supply voltages. Autotransformers are used also in RF amplifiers, in which they function as tuned transformers.

avalanche the uncontrollable breakdown of insulation, particularly in a gas or a SEMICONDUCTOR. An avalanche occurs when an ION is formed in a gas in which there is a strong electric field. The acceleration of the ion by the electric field can cause it to reach such speeds that collisions with gas molecules will generate more ions, which in turn are accelerated. This leads to total ionization of the gas in a short time. The initial ionization can be triggered by a charged particle from a radioactive material, as in a Geiger-Müller tube, or by light or other radiation. See also AVALANCHE BREAKDOWN.

avalanche breakdown a form of AVALANCHE effect in a SEMICONDUCTOR junction. Avalanche breakdown occurs in a reverse biased junction at a very sharply defined voltage level. The effect is used in the manufacture of voltage reference diodes. These diodes are usually named ZENER DIODES but ZENER BREAKDOWN is more gradual than avalanche breakdown.

AVC see AUTOMATIC GAIN CONTROL.

avionics the application of electronics to aviation. This covers all aspects of airborne radar, communications, control systems, and on-board computers.

azimuth angle the angle of tilt of a tape recording or replay HEAD. The head is at zero azimuth when the line of the magnetic gap is at 90 degrees to the line of motion of the tape. A tape recorded by a head with an azimuth angle error will replay poorly on a machine with a correctly aligned head. This is particularly important if the tape is used for digital signals. See Fig. 8.

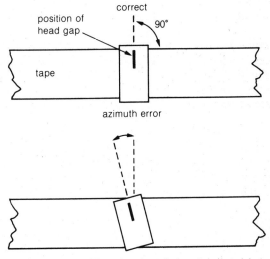

FIG. 8. **Azimuth angle.** The azimuth angle for a tape head. Incorrect setting of this angle is common and causes muffled sound when a tape created on another machine is played.

azimuth error an incorrect tape head AZIMUTH ANGLE that causes faint recording or replay.

B

---

back EMF a voltage set up in opposition to a supplied voltage, for example, the reverse ELECTROMOTIVE FORCE in an INDUCTOR when current changes, and the EMF developed in the ARMATURE coils of a DC motor when the armature turns.

background count the count of ionizing particles per minute in the absence of known radiating material. The background count is the natural count due to Earth's own radioactivity, and also to the bombardment of particles from outer space. Because this background is irregular and varies considerably from one place to another, careful measurements are needed to distinguish background from any additional radiation from a weak radioactive source. See also IONIZATION.

background noise the NOISE signal always present in any transmission.

back heating an effect used in a MAGNETRON to continue electron emission. Electron emission is started by heating the CATHODE of a magnetron. When the magnetron is running, however, the paths of many electrons cause them to return to the cathode. Bombardment of the cathode by returning (*back*) electrons causes heating, so the auxiliary heating can be reduced or switched off.

back lobe the pattern of strong radiation from a directional antenna that is in the reverse direction to the intended direction. See LOBE, RADIATION PATTERN.

back plate the SIGNAL ELECTRODE of a TV CAMERA TUBE. The term originally denoted a metal plate used as an electron collector and located at the rear of the tube. For modern types of camera pickup tubes, such as the vidicon, the back plate is a conducting layer at the front of the tube.

back porch the portion of the SYNCHRONIZING PULSE TV signal waveform that immediately follows the line sync pulse. On a monochrome signal, the back porch consists of 2.25 microseconds of BLACK LEVEL. On a color signal, the COLOR BURST of 10 cycles of subcarrier takes place in this time. This portion of subcarrier signal is used to synchronize the color subcarrier oscillator in the receiver.

backward diode a diode in which both regions are heavily doped (see DOPING). This has the effect of enabling the diode to conduct better in the reverse direction than in the forward direction, and with very small voltage drop. Because there is virtually no CARRIER STORAGE response speeds are high.

The backward diode is used for DEMODULATION particularly of low-level microwave signals. See also TUNNEL DIODE.

back-wave tube a form of TRAVELING-WAVE TUBE oscillator in which the electromagnetic wave travels in the opposite direction from the electron beam. See also CARCINOTRON.

balance the symmetry or equality of signals. Stereo audio signals are said to be balanced in a stereo amplifier if they produce a sound that appears to initiate centrally between the loudspeakers when both channels are fed with the same signals. A waveform is said to be balanced about ground if the shape of the wave form is symmetrical with zero volts as the axis of symmetry.

balanced amplifier an amplifier in which both a signal and its inverse are simultaneously amplified. The inputs and outputs of such an amplifier are balanced about ground (see BALANCE). The advantage is great immunity to interference from supply-line or radiated noise. See also PARAPHASE AMPLIFIER.

balanced line a two-wire transmission line that carries signals balanced with respect to ground (see BALANCE). The wires should have identical electrical characteristics.

ballast a resistor used in early radio and electronic circuits to absorb supply SURGES. A ballast resistor was wired in series with a DC supply. Because the resistor ran hot and had a large positive TEMPERATURE COEFFICIENT of resistance, any increase in current through the resistor caused an increase in its resistance value. This stabilized the value of current through the ballast. Tungsten-filament lamps were often used as ballast resistors.

balun *acronym for bal*anced-*un*balanced transformer, a transformer designed to match an unbalanced line to a balanced line with minimum power loss. Its main use is to connect a (balanced) dipole antenna to an (unbalanced) coaxial feeder.

band 1. a range of frequencies. 2. in semiconductor theory, a range of permitted energy levels (see ENERGY BANDS).

band energy see ENERGY BAND.

band-pass filter a filter that permits a band of frequencies to pass, but rejects frequencies above and below that range.

band-stop filter a filter that stops a range of frequencies, but permits signals below or above that range to pass.

band switch a manual switch for changing the tuning of a receiver or transmitter from one frequency range to another. This is usually done by changing the connections to inductors.

bandwidth the range of frequencies in a band. For radio frequencies the range is usually given in terms of the difference between the highest and the lowest frequencies in the band.

bank a set of controls arranged together. The controls—switches, potentiometers, variable capacitors—may be connected mechanically or electrically so all of them can be altered by one control.

barrage reception see DIVERSITY RECEPTION.

barrel distortion the distortion of a TV picture in which the normally rectangular shape is distorted to a shape in which the sides are bowed outward. Compare PINCUSHION DISTORTION.

base one terminal of a BIPOLAR TRANSISTOR. For maximum voltage gain, the signal input will be to the base, and the output from the COLLECTOR. In some applications, the base is held at AC zero potential, and the signal input is to the EMITTER (1).

base level the voltage level from which a PULSE begins and to which it returns. In most pulse-generating circuits the base level will either be ground or supply voltage level.

base stopper a resistor of low value that is wired in series with the base of a TRANSISTOR in order to damp PARASITIC OSCILLATIONS. The base STOPPER is generally wired as close to the base terminal as is physically possible.

BASIC *acronym for b*eginner's *a*ll-purpose *s*ymbolic *i*nstruction *c*ode, a high-level computing language, originally intended for teaching the skills of programming, but now useful and well known in its own right.

basic frequency see FUNDAMENTAL FREQUENCY.

bass the lowest pitched audio frequency notes in the region 15Hz to 200Hz approximately. These frequencies are attenuated by the normal capacitor coupling in amplifier circuits, and the lower bass frequencies are not well reproduced by loudspeakers, particularly by small loudspeakers.

bass boost a selective gain increase for the BASS frequencies. Bass boost is often used to compensate for the attenuation of bass frequencies in amplifier circuits and for the deficiencies of loudspeakers.

bass response the relative gain of a device for the BASS frequencies, usually given in decibels relative to the response for a 1kHz signal.

battery a group of identical devices connected so as to enhance one or more properties of an individual device. The term usually denotes a group of cells connected so as to provide either higher voltage or higher current than provided by a single cell. Series connection provides voltage levels equal to the cell voltage multiplied by the number of series cells. Parallel connection enables higher currents to be delivered, because the total internal resistance is equal to the internal resistance of a cell divided by the number of cells.

baud the unit of rate of serial transmission of digital signals along a pair of wires. One baud denotes one signal voltage or phase change per second. The unit is named for the pioneer telegraph engineer, J. Baudot (1845—1903). Calculating the rate of information transmitted at a given band rate is complicated by variations in the size of a BYTE and by the presence of start and stop BITS.

bayonet socket any form of locking connector that is attached by a plug-in and twist action.

beacon or **radio beacon** the radio equivalent of a lighthouse. Signals sent out from a beacon transmitter are used for identification of the transmitter (*code beacon*), for location and guidance (*homing beacon*), and for more specialized purposes such as aircraft instrument landing.

bead thermistor a form of THERMISTOR in the shape of a small sphere with two leads.

beam a confined path for radio waves or electrons. The term implies that the shape of the path is almost parallel-sided and straight.

beam angle the angle of divergence of electrons emerging from the ELECTRON GUN of a CATHODE-RAY TUBE.

beam bending a form of distortion in TV CAMERA TUBES. The electron beam scans areas that are electrostatically charged by the action of light. The discharge current provided by the beam then constitutes the output video signal. The differences in charge, however, can cause the low-velocity beam to be deflected in the direction of more positive charge. The effect on the signal of this beam bending is to exaggerate the size of bright areas in the picture.

beam lead a connection on a transistor or integrated circuit that is resistant to vibration. This is achieved by supporting the connection at the center so it is balanced.

beam-power tube a form of vacuum tube in which the electron flow from cathode to anode is focused into a beam by beam-forming plates. This results in lower losses than achieved through use of an additional grid.

beam switching a method of obtaining more than one TRACE from a single electron gun of a CATHODE-RAY TUBE. This is done by rapid alternation of vertical position of the trace so that to the eye more than one trace appears. The signals to the DEFLECTION PLATES are also switched at the same rate, making it appear that completely independent traces are being used.

beat frequency a signal produced as a result of mixing two other signals. When two signals, frequencies $f1$ and $f2$, are mixed in a NONLINEAR device, the output contains these frequencies along with the sum frequency ($f1 + f2$) and the difference frequency ($f1 - f2$, or $f2 - f1$, whichever is positive). These sum and difference frequencies are both beat frequencies, but the difference frequency is the one that is normally used and is referred to as a beat frequency.

beat-frequency oscillator (BFO) a part of a receiver for CW (*continuous wave*, meaning Morse code) signals that operates at a frequency slightly above or below the INTERMEDIATE FREQUENCY. The difference in frequencies is usually about 400Hz to 1kHz, so the beat frequency is an audio note. The BFO output is applied to the demodulator, so that Morse signals sound like audio notes in the audio frequency sections of the receiver.

The term can also denote an audio SIGNAL GENERATOR that makes use of two radio frequency oscillators that are mixed to produce a wider range of beat signals.

beating the production of signals from the mixing of higher frequencies.

beat note or **beat** a signal produced by BEATING.

bel a unit for comparing two power levels, equal to the logarithm to the base ten of the ratio of the two powers. *Abbrevs:* B, b. Named after Alexander Graham Bell, US scientist born in Scotland.

In electronics, power gain in bels is

$$\mathrm{Log}_{10} \frac{P_2{}^2}{P_1{}^2}$$

where P_1 and P_2 are input and output powers respectively. A more convenient unit is the DECIBEL.

beta gain the COMMON EMITTER current gain of a transistor.

beta ray see ELECTRON BEAM.

BFO see BEAT-FREQUENCY OSCILLATOR.

bias a DC voltage applied to a device in order to control its linearity (see LINEAR). The output-input characteristic of most active devices is nonlinear, and bias is used to select a position in the characteristic curve that will give reasonably linear results. Bias can also be used to provide deliberate nonlinearity, so the tip of a pulse, for example, can operate a circuit.

bidirectional transistor a transistor in which the COLLECTOR and EMITTER regions have identical doping levels and geometry, so the collector and emitter leads can be interchanged. Many small-signal transistors are reasonably bidirectional.

bifilar winding a method of coil winding that reduces stray INDUCTANCE losses. The usual technique is to wind both the primary and the secondary wires together around the core. See Fig. 9.

bimetallic strip a metal strip that will bend when its temperature changes. The strip is a composite of two metals with different

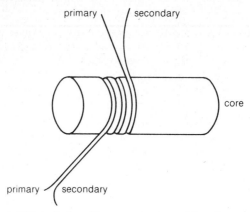

FIG. 9. A **Bifilar winding.** The separate wires of the primary and the secondary windings are wound together onto the core.

expansivity values. The bending of the strip is normally used as a method of switching for thermostats, temperature limit switches, and the like.

binary code a system of representing numbers with only the digits 0 and 1. The term can include systems such as Gray code and Excess-three code, but it normally denotes 8-4-2-1 binary code in which the position of a digit in a number indicates its power of 2 significance.

binary number a number expressed in BINARY CODE.

Biot-Savart law the relationship between the amount of current flowing between two points and the MAGNETIC FLUX DENSITY the flow creates. Formerly called *Ampère's law*, the Biot-Savart law is stated in a form that is physically impossible to achieve, and only formulas deduced from the law can be tested.

The Biot-Savart law can be explained more fully by referring to Fig. 10. (Note that symbols in italic type, for example, *I*, indicate a scalar; and that symbols in bold type, for example, **B**, indicate a vector.)

For any point P and any element *dl* of the current, the vector (cross) product $\mathbf{u}_T \times \mathbf{u}_R$ is perpendicular to the plane defined by the point P and the current *I*. The direction of the vector product

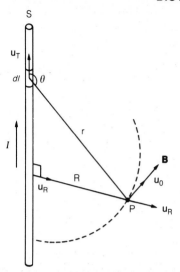

FIG. 10. **Biot-Savart law.** The law relates the flux density $\Delta \mathbf{B}$ to the current I flowing in a minute element of conductor ds long. The flux density is calculated at a radius R from the wire, and the angle between the conductor and a line drawn from the element to the point at which the flux is considered is θ.

is, therefore, that of the unit vector $\mathbf{u}_\theta$. The magnetic field produced at P by the element of current dl is a tangent to the circle of radius R that passes through P, is centered on the current, and is in a plane perpendicular to the current. The total magnetic field **B** is the sum of the contributions made by each element, that is,

$$\mathbf{B} = \frac{\mu_0}{4\pi} \, I \oint \frac{\mathbf{u}_T \times \mathbf{u}_R}{\mathbf{r}^2} dl$$

The contributions from all terms in the integral have direction $\mathbf{u}_\theta$, and the total field **B** is also a tangent to the circle at P. Therefore, it is only necessary to find the magnitude of **B**. As $\mathbf{u}_T$ and $\mathbf{u}_R$ are unit vectors, the magnitude of $\mathbf{u}_T \times \mathbf{u}_R$ is $\sin\theta$. We can rewrite the equation above as:

$$B = \frac{\mu_0 I}{4\pi} \int_{-\infty}^{\infty} \frac{\text{Sin}\theta}{r^2} \, dl$$

(where B is the magnitude of **B**), which can be solved to give the Biot-Savart law:

$$B = \frac{\mu_0 I}{2\pi R}$$

where R = radius.

bipolar IC a form of INTEGRATED CIRCUIT in which the transistors are bipolar (see BIPOLAR TRANSISTOR) rather than field effect types.

bipolar transistor a TRANSISTOR that employs two close-spaced JUNCTIONS(2) in a single CRYSTAL to form a p-n-p or n-p-n sandwich of layers. The central section is called the base, and the outer regions are the emitter and the collector. In general, the collector has a larger area than the emitter; otherwise, the transistor construction is symmetrical.

With a steady voltage applied between collector and emitter, no current can flow unless a current is also flowing between base and emitter. The collector-emitter current is much greater than the base-emitter current, but is proportional to the base-emitter current over a large current range. For the P-N-P TRANSISTOR, the controlled current consists of holes, and the collector and base voltages are negative with respect to the emitter. The N-P-N transistor uses electrons as carriers and operates with collector and base voltage levels that are positive with respect to the emitter. Compare FIELD-EFFECT TRANSISTOR.

bistable a circuit possessing two stable states. The term usually denotes a FLIP-FLOP circuit that can be switched from one stable state to the other by a TRIGGER pulse. Compare STABLE CIRCUIT; see also MULTIVIBRATOR, MONOSTABLE STEERING DIODE.

bit *created from bi*nary digi*t*, denoting one of the two possible digits, 0 or 1, in a BINARY NUMBER.

Bitter pattern a method of investigating the magnetism of a material. A liquid that contains fine magnetic particles is applied to the surface of a magnetic material. The particles then form into a pattern that reveals how the material is magnetized.

black box a method of treating the behavior of a circuit or other device in which the internal construction is ignored, and only the inputs and outputs are considered. By treating a device as a black box, its effect in the circuit can be studied rather than its internal action. Components such as INTEGRATED CIRCUITS generally have to be treated as black boxes, because their internal construction is known in detail only to their manufacturers.

black level a voltage level in a TV VIDEO SIGNAL that results in the electron beam being just cut off. Signals below this level cannot produce any effect on the screen of the cathode-ray tube.

blackout a sudden loss of circuit gain. This is normally the result of nonlinear action that charges capacitors to CUTOFF bias levels. The cause of the nonlinear action is usually an excessive pulse input.

blanking the suppression of an electron beam. This is normally done to prevent an unwanted display as, for example, when the FLYBACK part of a TIMEBASE is being traced.

blanking level the voltage level in a video signal that corresponds to electron beam cutoff. See also BLACK LEVEL.

blasting a severe distortion of an audio output caused by excessive input signal.

block diagram an overall diagram for a system. A block diagram shows no circuit details. Each part of the circuit is shown as a block, with drawings of the waveforms in and out to illustrate the actions. In this way, the overall action is easier to understand, and a fault can often be isolated to one particular block. See also BLACK BOX.

blocked impedance a form of IMPEDANCE found in an electromechanical TRANSDUCER. When a signal is applied to a loudspeaker or other electromechanical device, the measured impedance is greater than the electrical resistance of the coil windings. This is because of the power conversion to mechanical power. If the motion is prevented, the impedance changes to the lower value, which is called the blocked impedance. Compare MOTIONAL IMPEDANCE.

blocking capacitor a CAPACITOR whose main purpose is to pass

BLOCKING OSCILLATOR

signal frequencies between parts of a circuit that are at different DC levels without passing any DC current.

blocking oscillator a form of OSCILLATOR used for pulse generation. A blocking oscillator makes use of a capacitor in the BIAS circuit. After one cycle of high-frequency oscillation, the capacitor is charged and switches off the oscillator completely, so no further oscillation can occur until the charge has leaked away. The leakage is slow in comparison to the time of oscillation, so the circuit generates pulses with a low DUTY FACTOR. See Fig. 11. See also SQUEGGING.

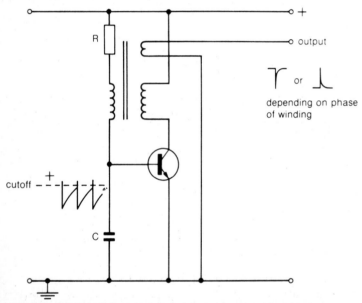

FIG. 11. **Blocking oscillator.** A typical blocking oscillator circuit. The time between pulses is determined by the values of R and C, and the pulse width by the amount of feedback and the characteristics of the transformer.

Bode diagram a DIAGRAM in which logarithmic gain and phase are plotted against logarithmic frequency for a filter, amplifier, or control system. The diagram is used to assess frequency response and stability, particularly for control systems.

body capacitance the CAPACITANCE of part of a circuit to a nearby human body. Body capacitance is particularly troublesome when variable capacitors are operated, because the setting obtained when the hand touches the adjustor may not be correct when the hand is removed.

bolometer a resistor for measuring radiated power, particularly infrared radiation, that is blackened so as to absorb radiation. When the radiation is absorbed, the resistor is heated, and its resistance value changes. The change in resistance value is then used to calculate the amount of power absorbed.

bond connect parts of a circuit so they are at the same ELECTRIC POTENTIAL usually ground potential. Bonding is reasonably straightforward for low-frequency circuits, but can be difficult for high radio-frequency circuits, when the length of the bonding conductor can be an appreciable fraction of a wavelength.

bonding pads the metal contacts around a semiconductor chip. The connections to the chip from each bonding pad are made through fine gold wires, and the bonding pads are in turn connected with more substantial wires to the pins of the chip.

booster 1. any device used to increase the power of a signal. **2.** a transmitter installed to increase signal strength in an area of poor reception. **3.** a carrier-frequency amplifier located at or near a receiver antenna to improve signal-to-noise ratio. **4.** a local power-supply transformer used to maintain voltage in a remote area.

bootstrapping a system of POSITIVE FEEDBACK across an amplifying stage whose gain is always less than unity, so that oscillation is not possible. Bootstrapping circuits are used to simulate constant current conditions in TIMEBASES and to increase the input impedance of amplifiers. See Fig. 12.

bottoming the DISTORTION of a signal because the signal voltage has reached ground level, or supply negative. An active device with a resistive load cannot provide an output voltage that rises above the level of the positive supply voltage or that falls below the level of the negative supply voltage, and distortion will be caused if the input signal is large enough to drive the output to these extremes.

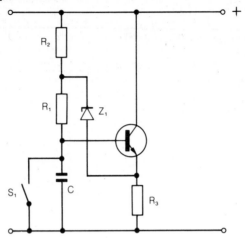

FIG. 12. **Bootstrapping.** A typical bootstrapping circuit for a time-base. When switch S1 opens, capacitor C begins to charge, and the voltage at the base of the transistor increases. Because the voltage at the emitter also rises and is coupled to the junction of R1 and R2 by the zener diode, the voltage across R1 is constant and the current through it is constant. This results in a linear rise of voltage across C, a ramp signal, until the voltage rise ceases or the switch closes. In a practical circuit, the switch would be a transistor.

branch a point at which three or more connections join, for example, a NETWORK of components.

breadboard a temporary assembly for testing purposes. Breadboard circuits are often constructed on plug-in boards, so circuits can be constructed simply by plugging in ICs, transistors, and passive components, with a minimum of wiring. See also PATCHBOARD.

break an interruption in a CIRCUIT.

breakdown a failure of insulation or other electrical characteristics. Breakdown of an active component generally means either loss of gain or a short circuit between electrodes.

bridge a form of measuring circuit. The simplest form of bridge circuit is the Wheatstone bridge, which consists of two potential dividers across a common supply (see Fig. 13). The condition for the potentials at the middle of the dividers to be identical is R1/R2 = R3/R4.

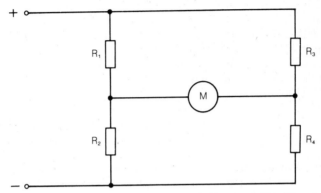

FIG. 13. **Bridge.** The Wheatstone-resistor bridge circuit. This form of the circuit diagram shows the principle more clearly than the traditional diamond-shaped pattern.

This condition, the balance condition, is easy to identify if a sensitive current indicator is connected between the points, bridging the gap. If three of the resistor values are known, then the fourth can be calculated. Alternatively, if one resistor in one arm is known, and the *ratio* of resistance values in the other arm is known, then the fourth resistance value can be calculated. The current indicator need not be calibrated, and its only requirement is that it be sensitive. The bridge, for example, can be supplied with AC and headphones or an amplifier used as a detector.

Variations in the basic bridge, using complex impedances in place of the potential dividers, can be used for measuring capacitance, self-inductance, mutual inductance, and transformer ratios.

bridge amplifier a circuit using two sets of output stages that provide signals, one of which is inverted. The load, usually a loudspeaker, is connected between the amplifier outputs. This is often used as a way of avoiding use of an isolating capacitor between an amplifier output and a loudspeaker, since the steady voltage across the output terminals can be set to zero.

bridge rectifier a full-wave rectifier circuit in which the rectifiers are connected in a bridge pattern, so a different pair of rectifiers conducts for each direction of AC input.

broadband or **wideband** (of amplifiers and networks) capable of operation over a large range of frequencies. Typical broadband systems include VIDEO AMPLIFIERS for television, and the signal circuits of OSCILLOSCOPES.

broadcasting the transmission of radio-frequency energy from an antenna so the signals can be picked up by any receiver. Radiated signals should be distinguished from signals sent along cables or optical fibers and delivered only to receivers connected to the cable.

broadside array a set of antennas arranged to receive or transmit in a direction at right angles to the line of the antennas. See also ANTENNA ARRAY.

brush a carbon, wire filament, or soft metal contact to a moving conductor, used on motors, servosystems, and rotary transformers.

brush discharge a faint glow discharge surrounding a high-voltage conductor. The brush discharge is particularly strong near points and sharp edges and causes severe radio-frequency interference.

BSI *abbreviation for* British Standards Institution, an association founded in 1901 that establishes and maintains standards for units of measurement, technical terminology, etc.

bubble memory a form of magnetic memory for computers using microscopic zones (bubbles) of reversible field inside a magnetic material to store information.

buffer 1. an isolating unity-gain amplifier. 2. a temporary memory in a computer.

buffer memory a part of computer memory used to gather data into convenient bundles. For example, data to be recorded on disk are often gathered into units of 256 bytes before being recorded. Replay from a disk may also be stored temporarily in a buffer.

bug a fault in a circuit or in a computer program.

build-up time the time needed for circuit current to reach its normal level.

buncher resonator the electrode of a KLYSTRON that changes the velocity of the electron beam in response to a signal.

buried layer a base layer of high-conductivity SEMICONDUCTOR. The layers of a transistor or integrated circuit are fabricated on top of this buried layer, which then is used for electrical and thermal connections.

burn-in see AGING.

burst a few cycles of signal. See COLOR BURST.

bus 1. a set of connections to several devices. A *power bus* for example, is a set of conductors that carries DC supply to several devices. 2. in computing, a set of lines that connects all the main components such as microprocessor, memory chips, and ports. See also ADDRESS, DATA BUS.

bush a form of bearing or insulator through which a cable shaft is run.

Butterworth filter a filter that produces a flat PASS BAND response, known as a *Butterworth response*, at the expense of steepness in the transition region between pass band and STOP BAND. The amplitude response is given by

$$\frac{V_{out}}{V_{in}} = \frac{1}{[1 + (f/f_c)^{2n}]^{\frac{1}{2}}}$$

where n is the order of the filter, f the frequency, and f_c the cutoff frequency. See also TCHEBYCHEFF FILTER.

by-pass (of a component) providing a shunt for signals. A *by-pass capacitor* for example, shunts unwanted signals to ground.

byte a group of (usually eight) BITS representing a single character.

C

cable an insulated wire or a set of insulated wires bound together (see INSULATOR). The individual wire CONDUCTORS in the cable are usually stranded, and the whole cable is reasonably flexible, so it can be laid into ducts. The term can denote a single SIGNAL conductor, such as the COAXIAL CABLE for a TV antenna, or a complex set of conductors like the cable that links a TV camera to its control unit. Most cables incorporate protection for the conductors, such as metal sheathing.

calibration the checking of a measuring instrument against a standard. The action is performed on instruments such as voltmeters and, particularly, CATHODE-RAY OSCILLOSCOPES whose characteristics may drift with time.

calibration pips a built-in form of timebase CALIBRATION for CATHODE-RAY OSCILLOSCOPES using short pulses generated by a crystal-controlled oscillator (see CRYSTAL CONTROL). By pressing a calibration switch, these calibration pips can be shown on the display, and used to check the accuracy of the timebase.

call sign a set of letters and numbers used to identify a transmitter. In the US and many other countries, even entertainment radio stations are required to broadcast a call sign at intervals.

camera tube a device that produces a VIDEO SIGNAL from an optical image. Camera tubes rely either on photoemissive or photoconductive effects to convert light intensity patterns into electrostatic charge patterns (see PHOTOELECTRIC EFFECT, PHOTOCONDUCTIVE CELL). A scanning electron beam then discharges these patterns, and as the beam scans, the discharging current forms the output video signal. A camera control unit provides the scanning signals for the camera and adds the correct BLACK LEVEL and SYNCHRONIZING PULSES to the video output so it can be used for studio monitors and ultimately transmitted. See also IMAGE ORTHICON, VIDICON.

cancellation circuit a technique used in radar circuitry to remove FIXED RETURNS. In a radar system that allows only moving objects to be noted, the pulses returned from fixed objects have to be canceled. This is done by using a DELAY LINE. The pulse from an object is compared with the pulse from the previous scan. If the pulses coincide, the object is nonmoving, and the cancellation circuit removes it by adding an inverted signal.

canned (of a portion of a circuit) being protected against electrostatic or magnetic fields. A SCREENING can placed around a coil reduces chances of pickup of interfering signals, and also reduces the field around the coil, which can affect other components. However, the can will lower the Q FACTOR of the coil.

capacitance the property of a system that enables it to store electrostatic charge. The capacitance of an isolated conductor is defined as the slope of the graph of potential plotted against charge. Symbol: C; unit: farad. An object is said to have a capacitance (C) of one farad if it can store one coulomb of charge (Q) at a potential of one volt.

Capacitance can be written as $C = Q/V$ and is measured in units of coulombs per volt. In practice the farad is too large a unit, so microfarads (10^{-6}), nanofarads (10^{-9}), and picofarads (10^{-12}) are used. See also CAPACITOR.

capacitive coupling a circuit that transfers signal through a CAPACITOR.

capacitive load a load that behaves like a combination of resistor and capacitor. This implies that the phase of voltage lags the phase of current. If the load were purely capacitive, the phase angle would be 90 degrees, but there would be no power dissipation.

capacitive reactance the ratio of signal voltage to current for a capacitor. This quantity decreases as the frequency is increased, with the relationship $X = 1/\omega C$ where X is reactance in ohms, ω is angular frequency (equal to $2\pi f$), and C is capacitance in farads. For a pure capacitor (with no series or parallel resistance and no inductance), the phase of current leads the phase of voltage by 90 degrees. See also IMPEDANCE.

capacitive tuning the altering of the tuning of a RESONANT CIR-

CUIT by varying the CAPACITANCE. This involves using a variable capacitor or a variable inductor.

capacitor a passive circuit component with CAPACITANCE. A capacitor is formed of a pair of conducting surfaces separated by a layer of insulator. A capacitor made of a pair of parallel conducting plates of area S separated by distance d with the gap between the plates filled by a dielectric of relative permittivity ϵ_r, will have a capacitance C given by

$$C = \frac{\epsilon_r \epsilon_o S}{d}$$

where ϵ_o is the permittivity of free space. The form of construction affects the amount of capacitance that can be achieved in a reasonable bulk, the stability of capacitance value, and the voltage that can be applied across the conductors (plates). The voltage-variable capacitance across a semiconductor junction can also be used in the form of a VARACTOR. See also CERAMIC CAPACITOR, ELECTROLYTIC CAPACITOR, FOIL CAPACITOR, MICA CAPACITOR, MOS CAPACITOR, PAPER CAPACITOR, PLASTIC-FILM CAPACITOR.

capacitor microphone a microphone in which the transducer is a CAPACITOR. The microphone consists of a metal backplate placed adjacent to a conducting diaphragm. The diaphragm is electrostatically charged, so when it vibrates, the voltage on the diaphragm will vary as capacitance varies. This variation of voltage constitutes the output signal.

The microphone tends to be bulky if reasonable sensitivity is required and needs a polarizing voltage to supply the charge to the diaphragm. The ELECTRET microphone is a more modern version.

capstan the rotating tape-drive spindle of a tape recorder. The capstan consists of an accurately ground cylindrical rod that rotates at constant speed. The tape is sandwiched between the capstan and a rubber pinch roller, or pinchwheel, so the constant speed rotation of the capstan causes a constant linear motion of the tape. Any irregularity of the capstan or of its bearings will cause irregular tape speed, which in turn will cause WOW or FLUTTER to appear on signals retrieved from the tape.

carbon microphone a microphone formerly used for telephones, now obsolete.

carbon resistor or **carbon-composition resistor** the cheapest and most common construction of resistor. It uses a mixture of graphite and clay, the same mixture used for pencils, baked and molded into a casing in contact with metal caps that are attached to the wire leads. The TOLERANCE range of resistance for a batch is large, and the stability of resistance value is not high. For most purposes, this type has now been replaced by metal or carbon FILM RESISTORS.

carcinotron or **crossed-field tube** a form of TRAVELING-WAVE TUBE that is used as a microwave oscillator. It is distinguished by the use of crossed electrostatic and magnetic fields, and a circular beam path. It is also classed as a form of BACK WAVE TUBE because the electromagnetic wave motion is in the reverse direction from that of the electron beam motion.

carrier 1. a radio-frequency sine wave that can be modulated (see MODULATION) and used to carry information. Also called *radio wave*. 2. an ELECTRON, HOLE or ION whose movement causes the flow of current.

carrier frequency the frequency of an unmodulated carrier wave.

carrier storage or **charge storage** the storage of charge in the form of excess CARRIERS near a semiconductor JUNCTION. The significance of carrier storage is that when the bias across the junction is reversed, current will flow as a result of the recombining of the carriers. This means that a junction will continue to pass current for a short time after applying a reverse bias.

carrier suppression a form of radio MODULATION. The carrier is modulated, giving a set of signals that consists of carrier and sidebands. The carrier is then suppressed by filtering or by adding an inverse carrier signal, leaving only the sideband(s). The effect of carrier suppression is to increase greatly the power efficiency of the transmission. Another effect is to confine reception to receivers that are equipped to replace the missing carrier.

cascade a number of devices connected in series so the output of one stage serves as the input of the next.

CASCODE

cascode a form of amplifier circuit of high output impedance and a large amount of isolation between output and input. In BIPOLAR TRANSISTOR form, the cascode consists of a COMMON EMITTER stage directly coupled to a COMMON BASE stage. See Fig. 14.

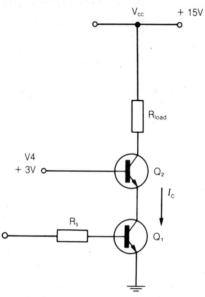

FIG. 14. **Cascode.** A cascode circuit using a pair of bipolar transistors. Q_1 is a grounded emitter amplifier with R_{load} as its collector resistor. Q_2 is interposed in the collector path to prevent Q_1's collector from swinging, while passing the collector current through the R_{load} unchanged. V_4 is a fixed-bias voltage. The cascode is an excellent HF amplifier.

catcher the electron-collecting electrode of a KLYSTRON.

catching diode or **clipping diode** a diode used to limit the amplitude of a waveform.

cathode or **emitter** the electron-emitting part of a VACUUM TUBE. In transmitting tubes, this consists of a winding of tungsten wire that is electrically heated to high temperature. An obsolete form of triode tube used a low-temperature cathode that consisted of a coating of oxides of calcium, strontium, and barium deposited onto a nickel tube. This was heated to a working temperature of about 1000 degrees K by a molybdenum filament inside the nickel tube. The filament was insulated by a coating of aluminum oxide (alumina). The cathode of a cathode-ray tube consists of a flat-topped nickel cylinder that is heated by an internal filament. The oxide coating is on the flat top of the cylinder.

cathode follower an obsolete form of vacuum tube circuit with high input impedance and low output impedance. The transistor equivalent is the EMITTER FOLLOWER or source follower. Today, follower circuits usually employ operational amplifiers.

cathode-ray oscilloscope (CRO) the instrument used for measurements of waveforms and pulses. Often shortened to *oscilloscope*. The oscilloscope is based on a CATHODE-RAY TUBE that uses ELECTROSTATIC DEFLECTION. The horizontal plates of the tube are supplied with a sawtooth TIMEBASE waveform that causes the electron beam to sweep across the face of the tube at a constant speed and return (fly back) rapidly. The beam is automatically brightened during sweep, and blanked during flyback. The waveform to be measured is amplified and applied to the Y-deflection plates.

The combination of waveform and timebase produces a graph of the wave, showing the amplitude and time between wave peaks. By having both the timebase and the amplifier calibrated, the peak-to-peak amplitude and the time between wave peaks can be measured for any wave that can be displayed. More advanced facilities include delayed triggering, second timebase, and multiple traces.

cathode-ray tube (CRT) an electron beam display device. All cathode-ray tubes make use of the emission of electrons from a hot CATHODE the formation of emitted electrons into a beam by ELECTRON LENSES and the emission of light from a material, the PHOSPHOR that is struck by electrons. In addition, all cathode-ray tubes must provide for the beam to be deflected in any direction across the face of the tube.

The two main types of cathode-ray tubes are INSTRUMENT TUBES used for measurement purposes, particularly in CATHODE-RAY OSCILLOSCOPES and PICTURE TUBES used for TV and radar.

cation a positively charged ION. The term 'cation' is used because a positively charged ion is attracted to a negatively charged electrode, or CATHODE. Compare ANION.

cavity resonator a MICROWAVE tuned circuit. The cylindrical cavity can be considered a coil with one turn that is tuned by stray capacity, but it is more realistic to regard it in terms of the dimensions of a wavelength as related to the circumference of the cavity.

CAZAMP *acronym for* commutating auto-zeroing amplifier, a device consisting of a pair of MOSFET OPERATIONAL AMPLIFIERS and a number of FET switches (see FIELD-EFFECT TRANSISTOR) contained within a single package. At any time, one amplifier is being zeroed, that is, having its input-offset voltage removed, while the second amplifier behaves like a normal operational amplifier and amplifies the signal. The operational amplifiers switch roles periodically via the FET switches, and the effect is that drifts due to input offsets are effectively canceled out.

CB see CITIZENS BAND.

CCD see CHARGE-COUPLED DEVICE.

CCIR *abbreviation for* Comité Consultatif International des Radiocommunications, the international standardizing committee for radio and television.

CCITT *abbreviation for* Comité Consultatif International Télégraphique et Téléphonique, the international standardizing committee for telegraphs and telephones.

cell any device that has an electric output. A *voltaic cell* is a generator of voltage from chemical effects, and such cells are used for low-voltage supplies, or when joined to make BATTERIES (1) for higher voltages. Light-sensitive cells, or PHOTOCELLS provide a voltage output when illuminated.

center tap a connection to the center of a transformer winding.

centimetric wave a wave whose wavelength is less than a meter and more than one centimeter.

central processing unit (CPU) the main working unit of a computer, which carries out the actions of the program.

ceramic capacitor a capacitor formed by metallizing two parallel surfaces of a ceramic slab. The range of CAPACITANCE can vary from a few picofarads to a few nanofarads. Ceramic capacitors often have capacitance values that can change as the applied voltage is altered, and can suffer from comparatively large energy losses. They are best used as bypass capacitors for radio frequencies, and when possible should be avoided in oscillator circuits.

cermet a ceramic-metal composite material, very hard but with a RESISTIVITY that makes it suitable for manufacturing POTENTIOMETERS.

channel 1. in radio communication, a specified frequency or frequency band used for signals. 2. in a FIELD-EFFECT TRANSISTOR the conducting path between the source and the drain whose conductivity is controlled by the gate.

channel stopper a heavily doped layer in an INTEGRATED CIRCUIT that prevents formation of unwanted FIELD-EFFECT TRANSISTOR channels.

characteristic a relationship or set of relationships between a number, usually two, quantities. These are usually shown graphically with one quantity plotted against another, and with a different graph line representing each value of a third quantity. See Fig. 15. For example, a set of characteristics for a TRANSISTOR may show the output current plotted against input current for a range of different output voltages.

characteristic impedance the impedance of a correctly terminated line or any other network with two input and two output terminals. See ITERATIVE IMPEDANCE.

charge the fundamental property of the electron (negative charge) and proton (positive charge). The recognizable and measurable feature of charge is that one charge will exert force on any other charge. The relationship was discovered by Charles Coulomb, for whom the modern unit of charge is named. One COULOMB, which equals 6.24184×10^{18} electrons, is the charge that, one meter from an equal charge in a vacuum, repels that charge with a force of 8.9874×10^9 newtons.

charge carrier any moving particle that is charged. The term denotes electrons, holes, and ions that move under the influence of an electric field. The movement of a charge carrier constitutes current. Current is defined as rate of flow of charge in units of coulombs per second, that is, $I = \mathrm{d}Q/\mathrm{d}t$. See also RESIDUAL CURRENT.

charge-coupled device (CCD) 1. a charge-storage memory de-

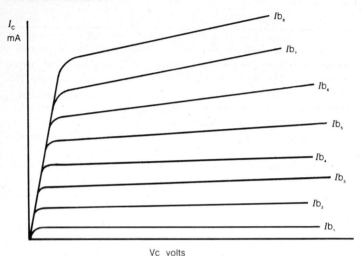

FIG. 15. **Characteristic.** A typical set of transistor characteristics. In this example, collector current I_c is plotted against collector voltage V_c, for various values I_b of base current.

vice. A CCD is based on a set of MOS capacitors formed in sequence on a chip, and with a common (ground) plate. By connecting adjacent groups of three capacitor plates to three-phase CLOCK signal lines, a charge placed on the first capacitor can be moved in sequence to each of the other plates on the chip. This forms a type of FIRST-IN/FIRST-OUT memory that is used mainly as a SHIFT REGISTER time delay, or serial memory. 2. a form of semiconductor TV camera device.

charge density a measure of the concentration of electric charge. The surface density of charge is measured in terms of the number of COULOMBS of charge per square meter of surface. A more practical unit is microcoulombs per square meter. The volume density of charge is measured in terms of coulombs or microcoulombs of charge per cubic meter of volume.

charge storage see CARRIER STORAGE.

charge-storage diode or **snap-off diode** or **stop-recovery diode** a diode that can be used as a pulse generator. The stored charge allows the diode to conduct for a brief time after applying reverse voltage. This causes the reverse voltage across the diode to remain low until all the charge carriers have been absorbed, following which the reverse voltage rise is very fast. This rapid voltage rise, limited by the CAPACITANCE across the diode and the supply impedance, constitutes the leading edge of a pulse that is terminated when the diode conducts again.

charge-storage tube see STORAGE TUBE.

charge-transfer device any form of INTEGRATED CIRCUIT that operates by switching charge from one capacitor to another. See also CHARGE-COUPLED DEVICE.

chart recorder an electromechanical device used for making a permanent record of a waveform. The input signal amplitude operates a pen drive so that the position of the pen on a paper chart indicates instantaneous signal amplitude. The chart may be in the form of a roll that is driven past the pen at a steady slow rate.

By using several separately controlled pens, a number of quantities can be plotted on one chart. For rapid measurements, photosensitive paper may be used with a point light source instead of a pen.

chip the small slice from a SEMICONDUCTOR wafer on which a transistor or integrated circuit is fabricated.

chip-select pin an input pin for a DIGITAL IC. The chip-select pin permits the complete chip to be enabled or disabled according to the logic voltage on the pin. For example, various areas of memory can be activated in a computer.

choke an obsolete term for INDUCTOR particularly a large inductor with a magnetic core.

choke coupling the use of an INDUCTOR as the load for an amplifying stage.

chopper a circuit that converts a steady input voltage by regular switching into a square wave. This was originally done mechanically, by means of a *vibrator*, but is now achieved with MOS transistors or integrated circuits. The chopped voltage can then be amplified, or stepped up, using AC amplifier techniques. DC-DC converters can be made by using a chopper, amplifier or transformer, and rectifier system.

chrominance signal the portion of a TV signal that carries color information. The TV signal has to permit compatibility, meaning that the same signal can be used by both MONOCHROME and COLOR TELEVISION receivers. The monochrome, or LUMINANCE signal is transmitted normally, and the color information is added in the form of a chrominance signal that is modulated onto a SUBCARRIER.

A monochrome receiver does not demodulate this subcarrier, and thus uses only the luminance signal. The effect of the subcarrier on the monochrome receiver is to cause a fine bar pattern that is not visible at normal viewing distances. A color receiver can demodulate the subcarrier, recover the chrominance signals, and make use of them, mixed with the luminance signal, to provide three separate signals for the three ELECTRON GUNS of the COLOR TUBE.

circuit a path for electric current, consisting of a current path through ACTIVE COMPONENTS or PASSIVE COMPONENTS. See also GAIN, FILTER, MODULATION, DEMODULATION, SWITCH, OSCILLATION and GATE.

circuit breaker a form of overload switch used in place of a FUSE to interrupt a circuit when excessive current flows or when a voltage surge occurs. The circuit breaker is magnetically operated and can be manually reset.

circuit diagram or **schematic** a diagram showing a circuit in symbolic form. There is a standard symbol for each component, showing the connections. In a circuit diagram, a current path is shown as a line that joins components. Some diagrams also show input and output waveforms at points in the circuit. Components in the diagram may be labeled with an identifying letter and number, such as R234 and C122, and usually also with their values. Compare WIRING DIAGRAM.

circuit element any one COMPONENT in a circuit.

Citizens Band (CB) a portion of the radio frequency range that is set aside for amateur users who do not hold an amateur license. The CB band is 26.965 MHz to 27.405 MHz.

clamp force a signal voltage level to a particular voltage at a selected time. Clamping is particularly important in a VIDEO waveform, because the 30% voltage level of the video signal is the BLACK LEVEL. If this is not at the correct voltage when applied to the ELECTRON GUN correct black level will not be achieved. At the time when the signal is at black level, a clamp provides a low-resistance path to the correct voltage level. The clamp is then released for the remainder of the signal.

clamping diode a diode used to prevent a signal from exceeding certain voltage limits. The diode is connected between the signal and a reference voltage so that if the signal exceeds the reference voltage, current flows through the diode and the signal voltage is clamped. More elaborate circuits can be devised, using a BRIDGE connection of diodes with the connecting path opened by using pulses to bias the diodes.

class A the linear operating mode of an amplifier. An amplifier is operating in class A if current flows in the amplifying device for each or any part of the input signal waveform. The amplifying device is never cut off and never bottomed (see BOTTOMING). The mode is used in particular for audio voltage amplifiers, some audio output stages, and some RF amplifiers (for frequency-modulated signals).

class AB the class of amplification (GAIN) in which the amplifying device can be cut off or bottomed (see BOTTOMING) for a small part of the signal waveform. The mode is used for some audio output stages and in RF amplifiers.

class B the class of amplification (GAIN) in which the amplifying device passes no current for half of the signal input wave. This system can be used for audio output if a matching device is used to handle the other half of the wave. It can also be used for RF amplification with a tuned load, because the natural oscillation of the tuned circuit will provide the missing section of output signal.

class C the class of amplification (GAIN) in which the amplifying device is cut off for more than half of the signal waveform. The mode is used for high-efficiency RF amplifiers in transmitters, CONTINUOUS WAVE (CW) or amplitude modulated. The load is always a tuned circuit whose resonance will supply the missing part of the waveform.

class D the class of amplification (GAIN) in which the amplifying device is based on a pulse-modulation amplification system. Pulses at high frequency have their width modulated by a low-frequency (usually audio) waveform. The pulses can be amplified with no regard to linearity, and demodulation is achieved by any low-pass filter, even by direct coupling to a loudspeaker. The system requires good pulse handling, and early examples were not competitive with conventional analog amplifiers.

C layer the lowest layer of the IONOSPHERE approximately 40 to 60km above the surface of Earth.

clear 1. reset a LOGIC output voltage to zero. **2.** empty a computer memory and fill it with null characters.

clipper or **peak limiter** a circuit that removes part of a waveform. A peak clipper is typically a biased diode circuit that will conduct when a signal peak exceeds the bias level. It can be used for limiting the amplitude of the wave peaks. Clippers are used extensively in speech-quality radio transmitter circuits to prevent OVERMODULATION. Another application is to remove peaks caused by the effects of IMPULSE NOISE on FM or TV signals. See Fig. 16.

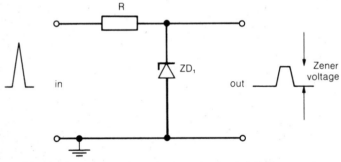

FIG. 16. **Clipper.** A simple peak clipper circuit using, in this example, a zener diode.

clock a circuit that generates pulses at precisely controlled intervals. Used in connection with logic gates and counters of the SYNCHRONOUS type, and particularly in computer circuits. Clock circuits generally consist of QUARTZ-CRYSTAL controlled oscillators along with pulse-shaping circuits.

closed circuit 1. a complete electrical circuit through which current can flow when a voltage is applied. **2.** closed, that is, wire-transmitted, applied to TV and radio systems.

clutter unwanted signals in a radar system. Clutter consists of signal echoes from such things as ocean waves, plowed fields, buildings, and other irregularities that are not radar targets. An *anticlutter circuit* is one that conceals such signals by making use of the fact that the signals always recur at the same times. By submitting a freshly received signal and a delayed signal to an ANTICOINCI-DENCE CIRCUIT the fixed signals of clutter can be removed, though this may also remove slowly changing signals from slowly moving targets.

CML see CURRENT-MODE LOGIC.

CMOS *abbreviation for* complementary metal oxide semiconductor, an INTEGRATED CIRCUIT structure that uses series combinations of n-channel and p-channel FIELD-EFFECT TRANSISTORS in each device. The important feature of CMOS construction is low power consumption, along with ability to operate on low supply voltages. For computing purposes, the older CMOS integrated circuits operate too slowly to be of use as processors, but CMOS memory with battery backup is sometimes used as a form of non-volatile memory, and modern CMOS designs are capable of faster operation. High-speed CMOS versions of microprocessors are available for small computers.

coaxial cable a type of shielded cable for conducting signals. The signal carrier is stranded or single-core wire that is insulated. This insulated sleeve is surrounded by a braided wire covering, form-ing the outer grounded conductor of the cable. This braiding in turn is covered by a tough outer insulator. Coaxial cable is used extensively in TV (RF and video) connections. The CHARACTERIS-TIC IMPEDANCE is normally about 50 to 75 ohms. By using solid inner wire, with insulating beads, the capacitance per meter and the loss per meter of the cable can be made very low. Compare TWIN CABLE.

coercivity or **coercive force** the amount of magnetic field strength that has to be applied to a magnetized material in order to reduce the flux density of the material to zero. When the field is removed, the magnetic material will have a remanent (see REM-ANENCE) flux density. See MAGNETIC HYSTERESIS.

coherent oscillator an oscillator whose bursts of waves are all in PHASE such as light and microwave oscillators of the LASER and MASER types respectively.

coil an INDUCTOR formed by a winding of wire or other conducting material into a shape with circular, square, elliptical, or other cross sections. The conductors must be insulated from each other, and the coil is used either for its INDUCTANCE or to generate a MAGNETIC FIELD.

coincidence circuit a circuit that delivers an output when two inputs coincide. A pulse coincidence circuit will deliver an output pulse when two input pulses to separate terminals arrive at the same time. A voltage coincidence circuit, or *voltage comparator*, provides an output when the voltage levels at the two inputs are identical, or within a preset range of each other.

coincidental half-current switching a method of using MAGNETIC CORES as computer memory elements. The magnetic characteristic graph shapes of the cores are approximately rectangular, and the magnetic state can be reversed by applying a known current. If each signal through one of a number of wires laced through the core consists of a current of about half the critical level, then a single signal cannot change the core magnetism. The magnetism will be changed only if two wires carry current signals in the same direction, hence the term "coincidental half-current switching."

cold-cathode tube an electron tube that emits electrons without having to be heated artificially. *Cold emission* of electrons is achieved by using a high value of electric field near the *cold cathode*. See also FIELD EMISSION.

cold emission see COLD-CATHODE TUBE.

collector the CARRIER collecting electrode of a transistor. Carriers that flow from the EMITTER (1) through the BASE region end up in the collector region. The total rate of flow of charge into this region constitutes the collector current.

color burst the 10 cycles of sine wave used to synchronize the color oscillator in a TV receiver. These cycles of sine wave are at the color SUBCARRIER frequency, and are placed in the BACK PORCH of the line sync signal.

color code the internationally standardized code used for marking resistors and, to a lesser extent, capacitor values.

color CRT a CATHODE-RAY TUBE that can display a range of colors. The most familiar type uses a set of three color-emitting PHOSPHORS that are laid down as fine vertical stripes on the face of the tube. Three separate electron guns are arranged in line, and a metal grille, the APERTURE GRILLE is placed close to the screen.

The effect of the apertures in the grille, one aperture for each group of three color phosphor stripes, is to shield each phosphor stripe from the electron beams of two of the electron guns. In this way, the beam from one gun can reach only the green phosphor, the beam from the second gun can reach only the red phosphor, and the beam from the third gun can reach only the blue phosphor. By controlling the relative beam currents of the three guns, any color and any brightness the system is capable of can be obtained.

color fringe a surrounding of unwanted color at the edge of an image. This can be caused by *moiré-pattern* effects, often called *strobing*, when a striped image is being viewed. It can also be caused by poor CONVERGENCE of electron guns, by undesired magnetization of tube electrodes, or by other faults. It can be particularly troublesome around monochrome images on COLOR TELEVISION tubes.

color killer a switching circuit within a COLOR TELEVISION receiver. The color killer switches off the color demodulator circuits when no color burst is being received. This avoids the possibility of random noise signals being interpreted as color signals and causing COLOR FRINGES on monochrome images.

color saturation a measurement of the intensity of color. A saturated color primary would be 100% red, green, or blue. Mixing white (itself a mixture of red, green, and blue) with a saturated color produces an unsaturated color. For example, a 50% saturated blue would consist of 50% blue and 50% white light.

color television the production of color images on a TV screen. The COLOR TELEVISION SYSTEM must be compatible, that is, capable of providing acceptable pictures on a MONOCHROME receiver. This is done by adding the color signals to the normal signal that would be used by a monochrome receiver. The signals from the camera are processed into a luminance signal, the normal signal for a monochrome receiver, and two difference signals.

A color difference signal is the signal voltage that results from subtracting the luminance signal from a color signal, and in the PAL system, these color difference signals are written as **R-Y** and **B-Y.** In this nomenclature, **Y** represents the luminance signal, **R** the red signal, and **B** the blue signal. These color difference signals are modulated onto a subcarrier that in turn is modulated onto the main vision signal. At the receiver, the color TV circuits will separate out the subcarrier signals, and generate a local subcarrier from an oscillator that is synchronized to the COLOR BURST in the BACK PORCH of the line sync pulse.

By using two synchronous demodulators (see SYNCHRONOUS DETECTION), the color difference signals are demodulated and then mixed with the luminance signal to provide three separate color signals—red, green, and blue. These are then used to operate the three electron beams of the color TV tube.

color television system the method used to carry the color difference signals to a COLOR TELEVISION receiver. The main systems are NTSC (US), PAL (most of Europe), and SECAM (France and USSR). The differences between the systems hinge on the methods of modulation of the subcarrier, and the way in which phase shift errors can be avoided.

Colpitts oscillator a crystal-controlled oscillator circuit whose distinguishing feature is a POTENTIAL DIVIDER that uses capacitors. See Fig. 17. See also HARTLEY OSCILLATOR.

common-base connection or **grounded-base circuit** a transistor connection in which the base is maintained at signal ground. The signal input is to the emitter, and the output is taken from the collector. The common-base connection offers unity current gain, moderate voltage gain, and good high-frequency characteristics. It is used mainly for impedance transformation (low input, high output impedances) and for radio frequency amplifier stages.

common-collector connection see EMITTER FOLLOWER.

common-emitter connection or **grounded-emitter connection** a transistor connection in which the emitter is maintained at signal ground. This connection features large current and voltage gain figures, with medium impedance input and output. It is the most common type of transistor circuit connection. See Fig. 18.

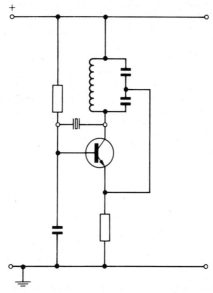

FIG. 17. **Colpitts oscillator.** One form of the Colpitts oscillator circuit, distinguished by its use of a capacitive voltage divider to obtain the feedback signal.

common-mode rejection a measure of the noise rejection of a differential amplifier. In such an amplifier, the difference between a pair of inputs is amplified. A common-mode signal is one that is applied with the same phase to both inputs of the amplifier, and it should result in a small output signal. The common-mode rejection ratio, expressed in decibels, is the ratio of the response to a differential signal to the response to a common-mode signal when both are applied at equal amplitude. The rejection ratio can be large, of the order of 100dB for some integrated circuits.

commutator a form of reversing switch, usually the mechanical rotary switch connected to coils in the armature of a motor.

compact disk a method of sound or video recording that uses a reflective or transparent plastic disk and a laser. The signals are converted into DIGITAL form and used to switch a laser beam on and off while the beam is being guided over the surface of a

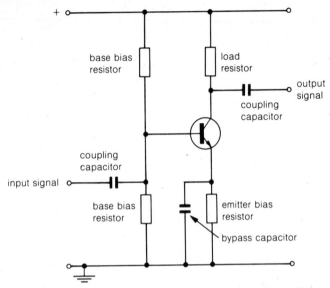

FIG. 18. **Common-emitter connection.** A typical common-emitter amplifier stage, shown here in the form of an audio-voltage amplifier.

spinning metal disk. This produces a pattern of pitting on the surface that carries the information of the signals, and this master disk can be used to print copies. The copies are called compact discs, and their signals can be played back with a system that uses a low-power laser. See also SOUND REPRODUCTION.

compander a combination of COMPRESSOR and EXPANDER. Many audio systems, notably recording systems, cannot cope with a large amplitude range, the DYNAMIC range of the signal. This can be remedied by using a NONLINEAR amplifier to compress the volume range before recording or transmitting. The volume range is restored later by another nonlinear amplifier, which expands the range again. The system is useful only if the compression and expansion can be matched closely.

comparator 1. (also called *voltage comparator*) a circuit based usually on an OPERATIONAL AMPLIFIER that gives a positive output when

$$V_{in1} > V_{in2}$$

a negative output when

$$V_{in1} < V_{in2}$$

and a zero output when

$$V_{in1} = V_{in2}$$

2. a device that compares two binary bytes and gives TRUE logical outputs to indicate whether one binary value is greater, less than, or equal to the other.

compensated attenuator an attenuator designed to operate over a large frequency range. A resistive attenuator does not provide the correct attenuation ratios for high-frequency signals. This is because of the STRAY CAPACITANCE across the resistors.

By using low value resistors, and by wiring capacitors in parallel with the resistors, which also provide the correct division ratio, the attenuator can be compensated for strays, and therefore be capable of operating over a wide bandwidth.

compensating leads a set of dummy leads used as balancing RESISTORS. In measurements that involve small changes in resistance values, the behavior of connecting leads may interfere with the measurement. For example, the leads may interfere because their resistance changes with temperature. By using a second identical pair of leads physically close, but electrically connected to the other part of the bridge, these changes can be compensated for.

complementary transistors a pair of transistors of opposite (bipolar) type, one a P-N-P TRANSISTOR the other an N-P-N TRANSISTOR or opposite CHANNEL MOS types. The two are often used in series circuits and must be closely matched in characteristics. The complementary bipolar pair is often used as an audio CLASS B output stage, and the MOS equivalent is the basis of CMOS integrated circuits. See also TOTEM-POLE CIRCUIT.

complementer a circuit whose output is the inverse of its input. See INVERTER.

COMPLEX NUMBER

complex number any number of the form a + bi, where a and b are real numbers and i = $\sqrt{-1}$ (engineers generally use j rather than i, because i may be mistaken for 1). Complex numbers include real and imaginary numbers. When describing AC circuits, it is necessary to use complex numbers to describe the magnitude and phase of a voltage or current.

A complex number can be represented as a point on a diagram, known as an *Argand diagram*, as shown in Fig. 19. Real numbers, such as DC voltages, appear as points along the x-axis. Complex numbers can appear anywhere on the diagram, and are defined in terms of a real part, a, and a complex part, b, in the form **c** = a + ib, where i = $\sqrt{-1}$, and **c** = complex quantity. The magnitude of **c** is $\sqrt{(a^2 + b^2)}$, and its phase is arc tan (b/a).

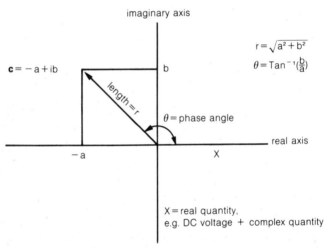

FIG. 19. **Complex number.** See this entry.

component any separately packaged unit of a CIRCUIT with its own connecting leads. See ACTIVE COMPONENT, PASSIVE COMPONENT.

composite conductor a multiwire conductor using different metals. One common type is a copper-steel composite, which is used for overhead power lines. The steel contributes little to the conductivity of the line, but most of its strength.

composite signal a VIDEO signal containing all the necessary components for a receiver. A video signal from a camera often consists only of the video portion from each color pick-up tube. The LUMINANCE signal has to be formed, and the color difference signals modulated on to the color SUBCARRIER. The SYNCHRONIZING PULSES are then added, along with the COLOR BURST to give the complete composite signal ready to be modulated onto the video carrier.

composition resistor see CARBON RESISTOR.

compressor a circuit that reduces the dynamic range of a signal. The action is achieved by a nonlinear stage whose gain reduces as the signal amplitude increases. See also COMPANDER.

computer any device that operates on data in accordance with an associated program of instructions. See ANALOG COMPUTER, DIGITAL COMPUTER.

conditionally stable (of an amplifier or servofeedback circuit) capable of oscillating only under certain conditions. This means the circuit will not break into oscillation unless some condition is violated. A common condition is that the load of an amplifier must be resistive or inductive, never capacitive. An amplifier that can be used with any load without any trace of oscillation is said to be *unconditionally stable.*

conductance the inverse of RESISTANCE. Symbol: $G;$ unit: SIEMENS. See also TRANSCONDUCTANCE.

conduction the transmission of energy by particle movement. Electrical conduction in most metals is the result of the movement of free electrons, but positive holes are present in significant numbers in some metals and are sometimes the main current carriers.

Conduction in gases and liquids is by movement of IONS. In liquids, mostly water solutions, the ions are surrounded by water molecules and move slowly. In gases, the ions can be accelerated rapidly so as to move fast enough to ionize any molecules they strike, thus causing total ionization. See also BREAKDOWN.

conduction angle the fraction of a cycle during which a device conducts, for example, a transistor operating in CLASS C. One cycle represents an angle of 360 degrees, and the fraction of the cycle for which the device conducts is multiplied by 360 to obtain the conduction angle.

conduction angle control a method of using a THYRISTOR to regulate an ALTERNATING CURRENT supply. In this system, the point in the cycle at which the thyristor fires can be controlled, and the thyristor is switched off by the zero-voltage part of each cycle. For a half-wave of AC, then, the thyristor can be on for a maximum of 180 degrees, and a minimum of zero. Full cycle control can be obtained using more than one thyristor, or by the use of a TRIAC.

conduction electrons the electrons in a solid that are free to move within the material, thus providing CONDUCTIVITY.

conductivity the inverse of RESISTIVITY. Symbol: κ; unit: siemens per meter. The conductivity of a material measures the ease with which electric current will pass through the material.

conductor a material of good CONDUCTIVITY for example, a metal or a semiconductor with a large density of free CARRIERS.

cone the normally conical part of a loudspeaker that vibrates to produce sound waves. The cone is made of stiff material. Materials such as paper, plastics, metal, and composites have been used at various times, but stiffened paper is the most widely used material. The cone is attached to the VOICE COIL, which is held by the SPIDER in the gap of the magnet.

conjugate (of impedance values) equal when the resistance portions are equal and the reactive components are equal and opposite, one inductive and the other capacitive.

constantan an alloy of nickel and copper used in manufacture of resistors that gets its name from having an almost constant resistance value as temperature changes. In other words, the TEMPERATURE COEFFICIENT of RESISTIVITY is unusually low. Since the resistivity value is fairly high for a metal alloy, fairly large value wire-wound resistors can be made of constantan without using long lengths or impracticably small diameters of wire. Used also in conjunction with pure copper in THERMOCOUPLES.

constant-current source a source of current that remains constant despite fluctuations in load. The main requirement for a constant current source is high output impedance. Thus, the current is determined by the source impedance rather than by the load impedance.

constant-K filter a type of FILTER design in which the product of series impedance and shunt impedance is constant and equal to the square of the characteristic resistance. See CHARACTERISTIC IMPEDANCE.

constant-voltage source a low impedance source that can maintain a constant voltage across a varying load.

contact a point or area where two materials touch. When the materials are both conductors, current can be made to flow across the contact. Unless the conductors are joined by high pressure, welding, or soldering, the resistance at this point, called the CONTACT RESISTANCE may be high. For two different metals in contact, there will usually be a potential of a few tenths of a volt DC that is called the *contact potential* or *contact voltage*. This contact potential is temperature-variable, so the contact can be used as a THERMOCOUPLE.

contact breaker an electromechanical protective device for CIRCUITS. The action of the contact breaker is to interrupt a supply when a fault condition arises, usually excessive current flow. Once the contact breaker operates, it has to be reset by hand to enable resumption of supply. See also OVERCURRENT TRIP.

contactor a form of relay for switching large currents. A contactor is designed with a larger mechanical movement than a relay, and with a parallel movement of contacts rather than the normal hinged-arm movement of a relay.

contact resistance the resistance between materials that are in CONTACT.

contact voltage see CONTACT.

continuous wave an unmodulated electromagnetic wave of constant amplitude and frequency, also known as *unmodulated carrier*. Information is impressed on the carrier when its amplitude, frequency, or phase is modulated. The simplest form of modulation is on-off keying, during which the carrier is varied, or keyed, between zero and maximum amplitude.

contrast the ratio of intensities of lit and unlit parts of a picture. The contrast of a TV signal is determined within limits by the

AMPLITUDE of the video signal. The contrast control of a TV receiver or monitor is the video GAIN CONTROL.

control electrode the electrode controlling current flow between other electrodes. The control electrode of a bipolar transistor is the BASE which controls current flow between emitter and collector. The controlling parameter in this example is base current.

For a MOSFET, the controlling electrode is the gate, and its voltage controls the current between source and drain. For a vacuum tube, the controlling electrode is the grid, the voltage of which controls current between cathode and anode. The grid voltage can be positive or negative with respect to the cathode voltage, and when the grid voltage is positive, grid current will flow.

convergence the meeting of the three electron beams of a COLOR TELEVISION tube. The three beams should converge and cross over at the APERTURE GRILLE.

conversion gain ratio a figure of merit for a FREQUENCY CHANGER equal to the ratio of output power at the changed frequency to input power at the original frequency. The figure is usually expressed in decibels.

converter any device that converts electrical energy from one form to another. An INVERTER (1) changes DC into AC. The circuit that changes from AC to DC is normally known as a POWER PACK.

The term can also denote a *frequency converter.* This is a circuit that beats signals together to form a new signal frequency. An *impedance converter* allows signal impedance to be matched to load IMPEDANCE. See COMMON-COLLECTOR CONNECTION, EMITTER FOLLOWER, and TRANSDUCERS.

cooling the dissipation of heat energy. For solid materials such as TRANSISTORS cooling relies first on conduction to transfer heat energy from the semiconductor to external materials such as cooling fins or metal blocks. The fins then, with a small amount of radiation, convect heat, and blocks can be cooled by water. Cooling for vacuum tubes can be by radiation or convection. Radiation-cooled tubes use a glass envelope, and the heat of the anode is allowed to radiate away. Larger tubes use metal envelopes that can be water cooled, steam cooled, or forced-air cooled.

coplanar transistor a form of TRANSISTOR construction in which the electrodes are all on the surface of the SEMICONDUCTOR.

copper a metal element universally used for low-resistance connections. Copper is used for wire, cable, and printed-circuit tracks.

copper-oxide rectifier a type of RECTIFIER that uses copper/copper-oxide JUNCTIONS. Formerly used in low-voltage supplies, particularly charges for secondary cells, until replaced by silicon. Copper-oxide rectifiers have a low forward voltage drop and are still used as meter-movement protection devices.

core see MAGNETIC CORE (1).

core loss the loss of energy from an inductor or transformer caused by the core. The two factors that cause loss are EDDY CURRENTS and HYSTERESIS. Eddy currents are caused by induction of voltages across parts of the core, which can be minimized by shaping the core, particularly with air gaps. Hysteresis loss is caused when the magnetism of the core is changed, and this can be minimized only by careful choice of material and operating conditions.

core store a computer memory system that uses MAGNETIC CORES. Each core, about 2mm in diameter, is threaded by perpendicular wires that magnetize the core in one direction or the other. The two directions of magnetization correspond to logic states 0 and 1. See COINCIDENTAL HALF-CURRENT SWITCHING.

core-type transformer a transformer wound onto a core so that the core is surrounded by the winding. The alternative is a *shell-type transformer*, in which the magnetic material surrounds the windings.

correlator a circuit enabling a waveform to be detected amid surrounding noise because of the periodic nature of the waveform.

cosmic noise the radio frequency noise that has its origins in space.

coulomb the unit of electric charge. Symbol: C. The coulomb is the charge that, when placed one meter from an equal charge in a vacuum, repels that equal charge with a force of 8.9874×10^9N. It is named after Charles Augustin de Coulomb (1736—1806), French physicist. The coulomb is an inconveniently large quantity for everyday use, so the microcoulomb (= 10^{-6}C; symbol: μC) and the picocoulomb (= 10^{-12}C; symbol: pC) are more commonly used. The coulomb can also be defined as the quantity of electricity transported in one second by a current of 1 ampere.

The definition of the coulomb is expressed in terms of Coulomb's law, which states that the force ($\mathbf{F}$) between two charges Q_1 and Q_2 at a separation of d meters in a vacuum is given by

$$\mathbf{F} = \frac{Q_1 Q_2}{4\pi\epsilon_\text{o} d^2}$$

where ϵ_o is the PERMITTIVITY OF FREE SPACE and F is in newtons.

counter a digital circuit giving an output that is the total of the number of pulses applied to the input. A digital counter may indicate the count on a digital display, and an analog counter may present it as a meter reading. Many counters display the COUNT RATE which is the number of pulses per second. Counters are used as frequency meters, and also to count the conduction pulses for detectors of radioactivity, such as GEIGER-MÜLLER TUBES.

countermeasures any electronic method of combatting enemy RADAR or communications or for counteracting JAMMING.

count rate the rate of arrival of pulses, expressed as pulses per second. See COUNTER.

coupling the passing of a signal from one stage to another. *Direct coupling* implies that the signal and its DC level are passed. *AC coupling* means that the DC is blocked, and only the alternating part of the signal is passed. *Tuned coupling* is used to pass one particular frequency or band of frequencies. The opposite is *broadband* or *aperiodic coupling.* See also CRITICAL COUPLING, INTERSTAGE COUPLING.

coupling coefficient a measure of the effectiveness of a tuned COUPLING. The coefficient is defined as the ratio of coupling impedance to the geometric average of impedances on each side of the coupling. The quantity is measured rather than calculated in most cases.

coupling loop a small inductor used to pick up a sample of a signal.

CPU see CENTRAL PROCESSING UNIT.

critical coupling the point of maximum transfer of SIGNAL between two RESONANT CIRCUITS. As the COUPLING between two tuned circuits is increased, the graph of amplitude plotted against frequency for the output rises in a steep peak at the resonant frequency (see RESONANCE). This peak is at its maximum at the point of critical coupling. See Fig. 20. See also OVERCOUPLING.

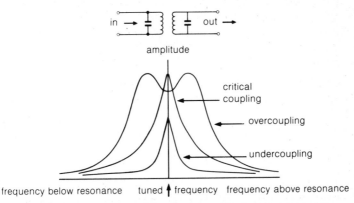

FIG. 20. **Critical coupling.** The effect of coupling between two tuned circuits, showing critical coupling, undercoupling, and overcoupling.

critical damping a resistive circuit loading just sufficient to prevent oscillation in a resonant circuit. If a resonant circuit is loaded by a resistor in parallel or in series, the resistor will dissipate energy and discourage oscillation.

The effectiveness of this resistor can be seen if the circuit is briefly pulsed and the output is viewed on an oscilloscope screen. If the damping is low, the circuit will oscillate freely, or ring (see RINGING). As the damping is increased (either a lower value of parallel resistance or a higher value of series resistance), the oscillations die away more quickly. Eventually, the shape of the output takes the form of a single cycle with no subsequent oscillation. This is the point of critical damping.

CRO see CATHODE-RAY OSCILLOSCOPE.

crosscoupling an undesired coupling between circuits, usually inductive circuits. Crosscoupling can be reduced by shielding and effective grounding. See also EARTH LOOP.

crossed-field tube see CARCINOTRON.

crossmodulation or **intermodulation** the unwanted modulation of one frequency by another. The effect occurs mainly in audio circuits and in wideband circuits in which a range of signal frequencies will be present. In audio circuits, crossmodulation causes noticeable distortion when the signal contains both a bass note and a treble note. Nonlinearity in an amplifier stage causes the bass note to modulate the treble note, so the sound from the loudspeaker gives the effect of a treble note whose loudness is rising and falling at the rate of the bass frequency.

crossover frequency a value of frequency that represents the transition between the passband and the stop band of a FILTER.

crossover network a network in loudspeaker systems. It is difficult to design a magnetic type of loudspeaker that will satisfactorily cover the entire audio-frequency range.

Because of this, high-quality loudspeaker systems contain a WOOFER a unit designed to give optimum reproduction of the lower notes, and a TWEETER, a unit or units that work best at high frequencies. The output from the amplifier then has to be divided so that the lower notes are fed to the woofer, not to the tweeter, and the higher notes are fed to the tweeter and not to the woofer.

The filter network that carries out this action is called the crossover network. At its simplest, it consists of a *capacitor* and a large-value *inductor*. See also DIVIDING NETWORK.

crosstalk interference between communications so that each channel receives some signal from the other. This can be due to unwanted coupling between telephone lines, between nearby amplifier circuits, or between different carrier frequencies. Crosstalk between different radio channels is often due to nonlinearity in some stage. This can occur when one transmitter is very close and is overloading the first stage of a receiver. The amount of crosstalk is usually specified in decibels.

CRT see CATHODE-RAY TUBE.

crystal a solid material in which the components, usually ions (see IONIZATION), are arranged in a precise geometrical pattern to create rigidity and hardness; the ability to be cleaved easily along certain directions, leaving a perfect plane surface; and in many cases useful electrical properties. QUARTZ CRYSTALS are used in oscillators, crystals of silicon for the manufacture of TRANSISTORS and INTEGRATED CIRCUITS and crystals of materials such as barium titanate in TRANSDUCERS.

crystal control the use of a vibrating QUARTZ CRYSTAL to control the frequency of a sine wave oscillator. The crystal acts as a tuned circuit with an exceptionally high Q-FACTOR (30,000 or more). This stabilizes the frequency of the oscillator and can also be used to produce a very pure sine wave, free from HARMONICS. For highest stability, the crystal must be cut in a certain way and contained in a thermostatically controlled enclosure, the crystal oven.

crystal counter a device for detecting IONIZING RADIATION. Several types of crystalline substances are affected when they are struck by radioactive particles. Some crystals emit a flash of light for each particle, and this light can be detected by a photocell, giving an electrical pulse. Another method is to separate an electron from a hole in a semiconducting crystal, which will make the material momentarily conductive. In either case, the resulting electrical pulse can be amplified and applied to a counter.

crystal filter a form of filter that uses a quartz crystal in place of a tuned circuit. This allows a narrow band of frequencies to be accepted or rejected.

crystal microphone a form of microphone that employs a PIEZOELECTRIC crystal. The sound waves vibrate the crystal, causing a varying voltage to be generated between conducting electrodes on the surface.

current the rate of flow of electric CHARGE. Symbol: I. Current is usually measured in AMPERES. The ampere is equivalent to the rate of flow of one COULOMB per second.

current amplifier an amplifier whose output signal current is much greater than its input signal current. If the voltage signal levels are unchanged, or if there is also voltage gain, the current amplifier acts also as a power amplifier.

current density 1. the amount of current flow per unit of cross-sectional area of a conductor. 2. a measure of radiated energy, such as from a radio beam.

current feedback the FEEDBACK of a signal that is proportional to the current output.

current gain the ratio of output current to input current for an amplifying device.

current limiter any device that will keep the current in a circuit from exceeding some fixed value independently of the voltage applied. A resistor is a simple form of current limiter, but the term is usually reserved for a circuit that is part of a constant voltage power supply. The function of the circuit is to limit the current to a safe level in the event that the load becomes a short circuit.

current-mode logic (CML) see EMITTER-COUPLED LOGIC.

current source/sink ability a measure in amperes or milliamperes, normally specified by the manufacturer, of the ability of a device to provide (source) or accept (sink) current without causing failure. See also FAN-IN, FAN-OUT.

current transformer a step-up TRANSFORMER whose low-impedance primary winding is connected in series with an AC supply line. The secondary voltage of the transformer is then proportional to the primary current. This secondary voltage can be rectified and used to drive a meter, an *instrument transformer*, that is scaled in terms of primary current.

A development of the idea uses a clamp as the primary winding. When the clamp is placed over an insulated cable, the meter reading shows the current passing. This is one of the few methods of measuring current without breaking the circuit.

cutoff the BIAS point at which current ceases in control devices such as CATHODE-RAY TUBES transistors, and vacuum tubes. The *cutoff voltage* of a CRT is the (negative) value of voltage between grid and cathode that will *just* cause the electron beam to disappear.

cutoff frequency 1. a value of frequency at which the gain of an amplifying device is noticeably reduced. 2. a frequency at the edge of the band in a band-pass filter at which the attenuation suddenly increases. This is usually taken as the point at which the signal is attenuated −3dB. 3. the frequency beyond which the gain of a transistor rapidly drops because of its construction. Once again, the −3dB point is usually selected.

cutoff voltage see CUTOFF.

cutout an automatic overload switch that will open under excessive current or dissipation conditions. The cutout may reset automatically after a time delay, or it may have to be reset manually. TV receivers often contain thermal cutouts that have to be reset by resoldering two spring-loaded contacts. A high-current cutout is called a CONTACT BREAKER.

cutter a device used in preparing a master recording on a disk. The sound signals from the master tape are applied to an electromagnetic cutter head that carries a stylus. The vibrating stylus cuts a groove of modulated width in the material of the master disk. This disk is then used to prepare a number of molds, and these in turn are used in stamping vinyl copies, the conventional audio records. The whole process is being superseded by the COMPACT DISK system (see RECORDING OF SOUND).

cycle a complete alternation of voltage or current. Typically, this may begin at zero, rise to a positive maximum, reverse, pass through zero, rise to a negative maximum, then return to zero again. A complete cycle is often represented by 360° or by 2π radians.

cylindrical winding or **solenoidal winding** $n.$ a coil wound on a cylindrical form. The coil can be single layer or multilayer, and if the length is much greater than the diameter, it is a not difficult to calculate the SELF-INDUCTANCE value in the absence of magnetic material. NOMOGRAMS are available for finding the number of turns of wire of a given gauge in order to achieve a stated inductance value.

D

daisychain connect (digital) devices so that a pulse will affect the first device in a series string that is enabled (see ENABLING), and that device in turn will affect the next in line.

An example is leading-zero suppression in a digital display so that, for example, the number 7.4 is not displayed as 00007.4. The zero-suppression pair of connections in each display unit is daisy-chained. When a number is displayed, a zero in the first display, the most significant digit, will be suppressed and a pulse will be sent to the next display. This process will continue until a display is reached that is not displaying a zero. The daisychaining ends there, and no other zeros will be suppressed.

damped (of a RESONANT CIRCUIT) prevented from oscillating. A resonant circuit is damped by a load of any type, such as a PARALLEL resistor. An undamped resonant circuit, which never exists in practice, would oscillate indefinitely after any stimulus. In practice, quartz crystals with no external damping can execute several thousand oscillations after being excited. Increased damping (a smaller value of parallel resistor) causes the amplitude of the oscillations to decrease. At critical damping, a pulse input produces a single cycle of output, with no subsequent oscillation. Damping can also be applied to mechanical oscillations, either mechanically, for example, using dashpots and oil, or electromagnetically. Compare OVERDAMPED, UNDERDAMPED, CRITICAL DAMPING.

damping factor a measurement of the effectiveness of damping (see DAMPED), that is, the ratio of the AMPLITUDE of an OSCILLATION to the amplitude of the following oscillation. The logarithm of this measurement, called the LOGARITHMIC DECREMENT, is also used.

dark current the current that flows in an unilluminated PHOTOCELL. This current, which can arise from molecular-scale defects in the photosensitive material or structure, limits the usefulness of photocells at low light levels.

dark spot a point on the PHOSPHOR of a CATHODE-RAY TUBE that is insensitive and does not shine when struck by electrons. A dark spot may be due to POISON in the phosphor, to ion bombardment, or to burn-out through having the beam current steadily on the spot over a long period.

Darlington pair a form of compound EMITTER FOLLOWER circuit for transistors. The circuit has high current gain, but its voltage gain is limited by the finite value of resistance between the collector and the base of the first transistor. See Fig. 21.

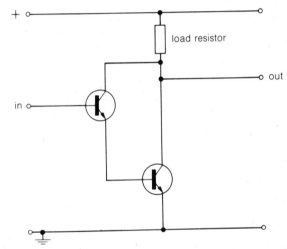

FIG. 21. **Darlington pair.** The circuit can be obtained in prefabricated form as a single device.

dashpot a mechanical damping device consisting of a small pot filled with oil. A cylinder floating in the oil is connected to the mechanical system whose motion is to be DAMPED.

data bus a set of connections carrying voltage levels representing binary digits (see BINARY CODE) to all parts of a computer system.

dB or **db** see DECIBEL.

dbx *Trademark.* a COMPANDER system. It is used for tape-hiss suppression on home cassette recorders and for increasing the signal-to-noise ratio for FM receivers.

DC see DIRECT CURRENT.

DC restoration the restoration of the correct DC level in a signal that has been transmitted by radio or passed through a capacitor. A VIDEO SIGNAL for example, that has been passed through a capacitor will not have its BLACK LEVEL at zero volts, and the voltage of the black level will vary with the ratio of light to dark in the signal. The DC position of the black level can be restored by clamping (see CLAMP), or more simply by a DC restoration circuit. See DC-RESTORING DIODE.

DC-restoring diode a diode used in a DC-RESTORATION circuit along with a capacitor to establish the correct BLACK LEVEL of a VIDEO signal. See Fig. 22.

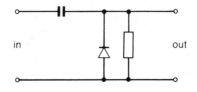

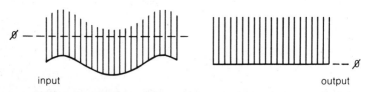

FIG. 22. **DC-restoring diode.** In this example, the principle is that the diode will conduct when the signal voltage is negative, charging the capacitor so as to restore a voltage level close to zero. The resistor ensures a time constant that will permit the capacitor voltage to be discharged slowly compared with the time between pulses. The forward voltage of the diode is ignored here. In practice it would cause the baseline of the output to be at about $-0.6V$.

deadbeat (of a meter movement) critically damped to avoid OSCIL-LATION. When the meter movement is connected in a circuit, the needle will not oscillate around a steady value, but will move steadily to that value. If the movement is OVERDAMPED the needle may take a long time to reach its final value.

dead time the time for which a device is disabled following some event. Many detectors of radioactivity are unable to detect an-other particle for a short time following the arrival of a particle, and this is the dead time for the detector. Some electronic circuits exhibit a dead time following an overload.

debouncing circuit a CIRCUIT used to suppress multiple pulses from a switch. When a mechanical switch is closed, the contacts will bounce open again, and may bounce several times. This causes closure of a switch to generate a set of voltage pulses. If the switch is part of a counter circuit, the excess pulses have to be suppressed, and this is done by a circuit connected across the switch. A typical circuit uses the RS FLIP-FLOP.

debunching the separation (due to mutual repulsion) of bunched electrons in a beam because of their CHARGE. Debunching affects dense electron beams, and particularly packets of electrons in KLYSTRONS.

deca- or **dec-** *prefix* denoting ten.

decade counter a counter that displays in the familiar scale of ten, as distinct from a BINARY CODE or HEXADECIMAL SCALE display.

decay time the time needed for a voltage to decay below a defined threshold. The decay time is often measured from the 90% to the 10% amplitude levels.

decibel (dB or **db)** the unit of comparative power, one-tenth of a BEL. For voltages V_1 and V_2 across equal impedances, the decibel ratio is defined as $20 \log_{10} V_2/V_1$.

decoder a demodulator for PULSE CODE MODULATION transmis-sions.

decoupling the process of shorting unwanted signals to ground. A decoupling CAPACITOR can be used to remove unwanted high frequencies, and an INDUCTOR can be used to decouple unwanted low frequencies.

decrement 1. the reduction of a count by one. 2. Also called *loga-*

rithmic decrement. a measurement of damping (see DAMPING FACTOR).

de-emphasis a circuit technique used in FM radio reception. The audio signal is pre-emphasized (see PRE-EMPHASIS) before modulation, and the deemphasis circuit is a form of LOW-PASS FILTER that restores the normal audio signal spectrum.

definition see RESOLUTION.

deflection defocusing a fault of an electron optical system resulting from the fact that a beam that is focused when aimed at the center of the screen or target will be out of focus at the edge. The cure is dynamic focusing, in which part of the deflection waveform is added to a focus current in a coil.

deflection plates a method of applying ELECTROSTATIC DEFLECTION to an instrument CATHODE-RAY TUBE. The plates are usually shaped in a diverging pattern to avoid being struck by the beam. The sensitivity of the tube will be proportional to the length of the plates.

deflection sensitivity a measure of the effectiveness of a deflection system. For ELECTROSTATIC DEFLECTION this is measured as the amount of beam deflection at the screen per volt applied between the deflection plates at a given acceleration voltage. The deflection sensitivity is inversely proportional to the accelerating potential between CATHODE and FINAL ANODE.

For a magnetically deflected tube, the deflection sensitivity is not internally fixed, because the deflection coils are external. The tube manufacturer may quote deflection sensitivity in terms of millimeters of deflection per unit of flux density. In this case, deflection sensitivity is inversely proportional to the square root of accelerating potential.

deflection yoke or **deflection coils** COILS used to supply magnetic fields for the BEAM deflection of a TV or radar CATHODE-RAY TUBE. The coils are shaped to wrap tightly around the neck of the tube and are fed with a sawtooth *current* waveform to cause beam deflection. The voltage waveform across the coils takes the form of a large pulse at the moment when current is switched off, and this is utilized in an energy recovery circuit, returning power to the supply and also generating the extra-high tension voltage for the FINAL ANODE of the tube.

defluxer a demagnetizing device. Most defluxers consist of a MAG-
NETIC CORE with a large air gap, and a coil that is connected to
AC supply. The large AC field can be applied to a magnetized
material to force the material to undergo the hysteresis cycle. If
the material is then pulled slowly away from the defluxing coils,
the magnetism will go through ever-decreasing peaks until it
reaches zero.

deformation voltage the voltage between the electrodes of a
PIEZOELECTRIC crystal when a force is applied to the crystal.

degree a unit of angle or of temperature.

delay a time interval or a voltage bias. The delay time for a pulse
circuit is the time needed for an input pulse to produce an output.
See also DELAYED AGC.

delayed AGC an AGC (AUTOMATIC GAIN CONTROL) voltage that
is applied against a BIAS. This causes the AGC to be ineffective
until the signal strength has reached some preset level. In this
way, small signals will benefit from the full gain of the RF and IF
stages of the receiver.

delay equalizer a network correcting the distortion of wideband
signals that have passed through a DELAY LINE.

delay line a circuit or system intended to introduce a time delay
in a signal. Delay lines can be made from COAXIAL or TWIN CA-
BLES. An alternative is to use a LUMPED PARAMETER line, consisting of a capacitor-inductor ladder network. For delays of more
than a fraction of a microsecond, ACOUSTIC DELAY LINES are more
practical. For digital circuits, CHARGE-COUPLED DEVICE delay
lines are used. See also CANCELLATION CIRCUIT.

delta-star transformer a form of TRANSFORMER for AC power
circuits. A three-phase supply network with no neutral line can be
transformed in this type of system to a three-phase and neutral
supply.

demodulation the recovery of a modulating waveform from a
modulated CARRIER. For AMPLITUDE MODULATION recovery can
be achieved by using a diode and capacitor to rectify the carrier.
For frequency modulation, the FOSTER-SEELEY DISCRIMINATOR
or RATIO DETECTOR has been used, but integrated circuits that use
pulse counting techniques are now more common. Phase sensitive
SYNCHRONOUS DEMODULATION is used for color TV CHROMI-

NANCE SIGNALS and various methods are used for the many forms of pulse modulation, ranging from a simple low-pass filter to elaborate decoders.

demodulator any circuit that accomplishes DEMODULATION. Formerly called a *detector*.

demultiplexer a circuit that separates signals that have been transmitted in combined form by a MULTIPLEXER. A *digital multiplexer* is a digital circuit that accepts binary signal inputs on a number of lines. Its outputs are a number of single lines. These output lines are numbered, and the one that is at low voltage (a logical FALSE) corresponds to the binary number on the input lines. The remaining outputs are at higher voltage (logical TRUE).

depletion layer the region in a SEMICONDUCTOR around a JUNCTION. In the region close to a P-N JUNCTION the number of free carriers is lower than in the remainder of the material, because they combine and eliminate each other. This is the depletion layer. Its size can be increased by reverse biasing the junction, and can be decreased by forward biasing. See also VARACTOR.

depletion mode the operation of a FIELD-EFFECT TRANSISTOR with reverse bias. A depletion mode FET will pass maximum current at zero bias, and the effect of bias is to reduce the channel current. Compare ENHANCEMENT MODE.

deposition the laying down of one material on top of another. See METALIZING.

depth of modulation or **modulation factor** a factor for AMPLITUDE MODULATION of a carrier that is equal to the amplitude of the modulating signal divided by the amplitude of the unmodulated signal.

Depth of modulation is usually expressed as a percentage and is an index of the efficiency of the modulating process. When the depth of modulation approaches 100%, the distortion of the signal will be very large, but when the depth of modulation is about 20% or less, the SIGNAL-TO-NOISE RATIO will be poor.

derating a reduction in operating RATINGS under some conditions. For example, the maximum voltage that can be applied to a transistor may have to be reduced when the temperature is increased or when high-frequency signals are used.

desoldering the removal of soldered connections. This can be done using solder braid, which removes solder by capillary action. A more suitable method for extensive use is a desoldering gun, which sucks up molten solder, using either a built-in or a separate hot iron to melt the solder.

Desoldering is particularly important if working integrated circuits are to be removed. For faulty ICs, it is easier to cut all the pins and remove them with a soldering iron and pliers. It is considered good practice to mount all ICs in sockets, so they can be removed easily if necessary.

destructive read a reading of memory that destroys the contents of the memory. MAGNETIC CORE (2) stores are subject to destructive read because the reading is carried out by pulses that will set the direction of magnetization to the logic 0 direction. This causes no effect on a read wire when the core is already at the 0 condition, but will induce a pulse when the core is at logic 1 and then is changed to logic 0. The core must then be rewritten to logic 1 by another circuit to prevent loss of the content of the memory.

detached contact system a method of indicating RELAY action in CIRCUIT DIAGRAMS. The coil of the relay is shown in the part of the circuit where it is wired. The contacts are then shown in whatever circuit they switch, and these may be illustrated on a separate page of the circuit. The correspondence of the coil and its contacts is shown by label numbers. See Fig. 23.

detector 1. an old term for DEMODULATOR. 2. a device that detects subatomic particles.

detune adjust a tuned circuit so that it is not in resonance with an input signal.

deviation 1. in error measurement and assessment, the difference from a standard value. 2. in FREQUENCY MODULATION the difference between the modulated frequency and the carrier center frequency at any instant.

device an electronic COMPONENT or small group of components that carries out an action.

diac a form of DIODE that conducts in either direction. The diac is an insulator for low voltages in either direction, but it will break down for a critical voltage and can pass currents in either direction. Diacs are used mainly in the triggering circuits of TRIACS and THYRISTORS.

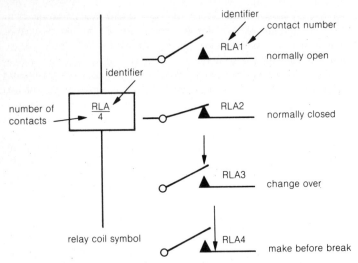

FIG. 23. **Detached contact system.** The figure shows the relay coil and its contacts separately, using identification letters and numbers so they can be associated. This system makes it relatively easy to sketch circuits containing relays.

diaphragm a thin vibrating sheet used in MICROPHONES to convert a sound wave into a mechanical movement that will then be converted into an electrical signal.

dichroic mirror a color-selective mirror consisting of a thin film of material on a flat glass plate. The thin film must have a refractive index different from that of glass. The thickness of the film is arranged to make use of interference effects so the film acts as a mirror for only a selected range of wavelengths.

dielectric an INSULATOR particularly in a CAPACITOR. The PERMITTIVITY of an insulating material was formerly known as the *dielectric constant.*

dielectric heating the process of heating an INSULATOR by making it the DIELECTRIC in a capacitor that is supplied with high-frequency signals. Heat is generated because of a form of electric hysteresis. See also INDUCTION FURNACE.

differential amplifier or (formerly) **long-tailed pair** an AMPLI-

FIER that has two inputs. The output of a differential amplifier is proportional to the voltage or current difference between the inputs. See also COMMON-MODE REJECTION.

differential capacitor a CAPACITOR with two sets of static plates and one set of moving plates. The moving plates can be adjusted so that their capacitance to either set of fixed plates is increased or decreased.

differential resistance see INCREMENTAL RESISTANCE.

differential winding a winding on a TRANSFORMER that can be used as a form of control by passing a current that opposes the current in the main coil.

differentiating circuit a CIRCUIT based on an OPERATIONAL AMPLIFIER with capacitive input and resistive feedback that carries out the mathematical operation of differentiation, that is, $V_{out} = dV_{in}/dt$.

When a voltage step is applied to a differentiating circuit, the output of the circuit is the differential of a function that is an impulse. The effect of a differentiating circuit on any waveform apart from a sine wave is to produce a signal with sharp spikes whenever the input waveform changes rapidly. The effect on a sine wave is to produce a cosine wave.

diffused junction a semiconductor JUNCTION (2) produced by DIFFUSION.

diffusion a method of creating a semiconductor JUNCTION by heating the pure semiconductor material in contact with a DOPING impurity. The doping impurity is in the form of a gaseous compound, and the semiconductor is heated to a temperature just below the melting point. At this temperature, atoms of the doping material that strike the surface of the semiconductor are able to diffuse into the semiconductor, forming a junction.

digital (of an electronic component or circuit) operating with a small number of voltage levels, usually two. When only two levels are used, they are described, no matter what voltage they correspond to, as level 0 (FALSE) and level 1 (TRUE). Because only two easily distinguished voltage levels are used, most of the forms of distortion that affect analog waveforms become irrelevant.

digital circuit a circuit whose inputs and outputs can take only two levels, referred to as (logic) 0 and (logic) 1. The word 'logic' is often omitted.

digital computer a computer that acts on numbers in BINARY CODE. Any type of information can be coded into number form, and the digital computer can carry out actions of selecting, storing, retrieving, and arranging the coded information. A particular advantage of the digital computer is that the programming can also be in number codes, so no circuit adjustment is required. ANALOG COMPUTERS require HARDWARE alteration.

digital counter a CIRCUIT that counts the number of pulses that arrive at its input. The counting is achieved using digital circuits, usually FLIP-FLOPS, and may be displayed in binary or denary figures, or even as a meter reading. The input pulse must be of a level that will operate the counter.

digital voltmeter (DVM) a voltmeter that measures a voltage level by counting the number of small voltage steps needed to reach that level. The digital voltmeter generates a waveform that consists of a STAIRCASE VOLTAGE wave, typically of one-millivolt steps. The steps are counted in a COUNTER and the count stops and is displayed when the staircase waveform voltage becomes equal to the applied voltage. The sequence automatically repeats several times a second so that varying voltages can be tracked.

digitizer any device that can convert a measurement into the form of a binary number. The easiest methods usually involve a TRANSDUCER to convert the quantity being measured into a voltage, and then a circuit like a DIGITAL VOLTMETER to obtain a binary number count.

DIL (dual-in-line) package a method of packaging an INTEGRATED CIRCUIT by attaching the chip to a plastic or ceramic slab that has a set of pins at 0.1 inch spacing along each side. The standard distance between the lines of pins is 0.3″ for the smaller packages, and 0.6″ for the larger types, such as microprocessors.

DIN *acronym for* Deutsche Industrie Norm, the German equivalent of ANSI, the standard applied to a system of plugs, sockets, and cables widely used for audio signals.

diode a COMPONENT that allows current to pass in only one direction between its two connectors. The direction in which current passes is the forward direction, and the voltage that causes current flow is the forward voltage.

A reverse voltage does not cause current to flow unless the voltage is high enough to cause diode BREAKDOWN. Virtually all diodes in use today are semiconductor diodes, using either point contact or junction construction. See also GUNN DIODE, IMPATT DIODE, TUNNEL DIODE, ZENER DIODE, LIGHT-EMITTING DIODE, PHOTODIODE.

diode forward voltage the VOLTAGE across a DIODE when it is conducting in the normal forward direction. This voltage is decided mainly by the semiconductor material from which the diode is constructed. Typical figures are 0.55V for a silicon diode, and 0.15V for a germanium diode.

diode pump a form of pulse-counting CIRCUIT that produces an output voltage proportional to the number of input pulses.

diode reverse voltage a VOLTAGE applied across a DIODE in the reverse, nonconducting direction.

diplexer a CIRCUIT that enables transmission of two signals along one wire or channel without interference.

dipole 1. any system of two equal and opposite elements such as a pair of equal and opposite electric CHARGES. **2.** see DIPOLE ANTENNA.

dipole antenna or **dipole** an antenna consisting of two conducting rods arranged in line but electrically separate. The rods are of equal length, cut to a specified fraction of the wavelength of the signal to be received or transmitted. A common size is half a wavelength, meaning half of the wavelength on the conductor rather than half of the free-space wavelength. For the lower range of frequencies, the dipole is more likely to consist of wires, possibly parallel sets of wires, rather than rods. The signals are applied to, or received from, the dipole halves in ANTIPHASE. See Fig. 24.

dip soldering a mass-production SOLDERING method. PRINTED CIRCUIT BOARDS have their components inserted, and the solder side of the board is coated with flux. The entire surface of the board is then dipped into a bath of molten solder. Solder is not allowed to run onto the component side of the board, but it comes in contact with all the joints on the underside. Where printed circuits are densely packed, solder repelling varnish is coated be-

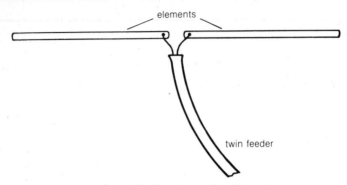

elements

twin feeder

FIG. 24. **Dipole antenna.** See this entry.

tween lines to prevent formation of solder bridges that would cause short circuits.

direct coupling the connecting of the output of one amplifier STAGE directly to the input of another. Direct coupling implies no DC blocking components, such as capacitors or transformers, so the DC level is also being amplified. Some designs attenuate the DC level by using resistors or ZENER DIODES in the coupling.

direct current (DC) a steady current in one direction such as is provided by a CELL a dynamo, or from the smoothed output of a rectifier circuit.

directional antenna an antenna that transmits or receives signals in one preferred direction. A simple DIPOLE ANTENNA is not particularly directional, but it can be made so by adding REFLECTOR and DIRECTOR elements. See Fig. 25. See also YAGI ANTENNA.

direction finding the determination of the direction of the transmitter of a radio frequency SIGNAL. The oldest method, *loop direction finding*, involves the use of two antennas in combination, one vertical whip and one rotating loop. A rotating loop antenna will give maximum signal when its axis points to the source of signal, but this occurs at two positions of the loop. By comparing the phase of the loop signal with a signal from a fixed antenna, the correct direction of the two can be found.

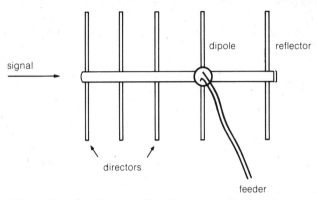

FIG. 25. **Directional antenna.** The directors of a directional antenna are located between the signal source of a receiving antenna and the dipole. The arrangement of the reflectors is also shown here. The entire configuration constitutes a YAGI ANTENNA.

Modern methods make use of precisely timed signals sent to and from satellites, and do not involve the geometrical arrangement of antennas.

directivity the ability of an ANTENNA system to transmit or receive in one particular direction.

director a rod element added to a DIPOLE ANTENNA. Director rods are placed in the desired direction of transmission or reception. They consist of conducting rods, but are not electrically connected to the dipole or to any signal. The distance of each director from the dipole or from the previous director as well as the overall length of the director must be calculated theoretically with a formula.

direct wave the wave path from TRANSMITTER to RECEIVER along Earth's surface. A direct wave is possible only if the transmitter and receiver are reasonably close, approximating to line of sight conditions. The wave received as a result of reflection from the ionosphere is known as the SKY WAVE.

discharge the removal or neutralization of electric CHARGE. An electrostatic charge on a material can be discharged by connection to ground or by IONIZATION of the surrounding air. An ionized gas passes current by movement of oppositely charged ions in opposite directions, and this is called a GAS DISCHARGE.

discrete (of a device) forming a separate component rather than part of an INTEGRATED CIRCUIT.

discriminator 1. a circuit that demodulates a frequency-modulated or phase-modulated signal (see DEMODULATION). 2. a circuit that selectively passes pulse inputs of some particular amplitude range.

disk 1. a flat plastic disk for recording sound. The conventional LP disk is made of vinyl and is hot-stamped with an impression of the sound pattern. The COMPACT DISK uses a form of digital recording and requires a laser-reading system. See also RECORDING OF SOUND.

2. a thin flat circular plastic sheet coated with magnetic material for the storage of computer data. See also FLOPPY DISK.

disk winding an old method of coil winding in which the coil is wound as a flat disk. Such coils can be seen on radios of the later 1920s and early 1930s.

displacement current the CURRENT that flows as a result of charge displacement in an INSULATOR. Displacement current is responsible for the REACTANCE of a capacitor and for TRANSMISSION of electromagnetic waves in space.

display any device that presents information as an image. Thus, CATHODE-RAY TUBES and digital SEVEN-SEGMENT DISPLAY devices are displays, but meter readings are not.

dissipation the loss of power from a system, either in the form of heat or as electromagnetic radiation.

distortion an unwanted change of signal waveshape as, for example, on a received TV picture. See also CROSSOVER NETWORK, DELAY, DEVIATION, FREQUENCY, HARMONIC, INTERMODULATION, NONLINEAR, PHASE DISTORTION.

distributed capacitance the CAPACITANCE that exists between conductors rather than contained in a capacitor. A pair of transmission lines will have capacitance to one another. This value is not located in one place or as one component, but is present distributed along the length of the line. For this reason, the value of capacitance is often given per meter of length. Similarly, the windings of a coil have capacitance to each other that is distributed rather than lumped (see LUMPED PARAMETER).

distributed inductance self-INDUCTANCE or mutual inductance that cannot be localized in a component. A conductor will have a value of inductance, for example, that is distributed along its whole length. The inductance is often given per meter.

disturbed-sun noise the additional radio frequency noise that results from sunspots or other eruptions on the sun.

divergence angle the angle between the sides of a BEAM. A beam of electrons that is required to have parallel sides will usually diverge because of the repulsion between electrons.

diversity reception or **barrage reception** a system of reception that uses several antennas or channels to carry the same information. The principle is to overcome interference or fading by using the strongest of the signals at any given time.

The most common type is in the form of a multiple-diversity ANTENNA system, using many antennas spread over a large area. The receiver then automatically selects the one providing the strongest signal at the desired frequency. Another form of diversity is *frequency diversity* which uses transmissions on several different frequency bands, with the receiver switched to the strongest signal at any time.

divider a CIRCUIT that carries out a reduction of a circuit quantity. A *potential divider* typically uses two resistors to reduce a voltage by a calculable amount. A *frequency divider* changes an input frequency to a lower value that is an exact submultiple of the original (half, third, tenth, etc.). Modern frequency divider circuits make use of digital systems.

dividing network a type of DIVIDER that separates signals on the basis of frequency. It will use potential divider circuits that contain reactances rather than pure resistors. See also CROSSOVER NETWORK.

division ratio the ratio of output signal to input signal for an ATTENUATOR or POTENTIOMETER.

D layer the layer in the IONOSPHERE that is located approximately between 60 and 90km above Earth's surface. This layer reflects mainly the lower radio frequency signals.

Dolby *Trademark.* a well known and widely used noise-reduction system for tape, particularly cassette tape. The principle is based on the fact that TAPE HISS is much more noticeable on low amplitude outputs. These are selectively boosted on recording and attenuated on replay. The system is therefore a companding system (see COMPANDER).

For professional use, the Dolby A system breaks the audio waveband into several sections and carries out the companding action for each section. The simpler Dolby B system, which is widely used for prerecorded tapes, confines the Dolby action to the higher frequency signals only. The later Dolby C system for home use is more effective and covers a wider range.

donor an impurity material that provides free electrons to a semiconductor. Compare ACCEPTOR.

doping the adding of a DONOR or ACCEPTOR impurity to a pure semiconductor.

Doppler effect originally the apparent change of sound frequency when either the source of sound, the receiver, or both are in motion. In sound, Doppler effect can be due to an increase of sound wave velocity when the source is moving. Electromagnetic waves travel at a constant rate in free space, and the Doppler effect for electromagnetic waves is due to changes in the wavelength of the signal caused by movement of source or receiver.

Doppler radar a RADAR system that relies on the change of echo frequency caused by a moving target. This is a system that is used for low-level navigation, and also in the radar speed tracking systems used by police.

dot generator a form of signal generator used for COLOR TELEVISION. The correct adjustment of a CATHODE-RAY TUBE particularly a color tube, is checked by connecting a dot generator to the receiver. This produces a pattern of dots that should be white and evenly spaced. Any irregularity in the spacing of the dots points to timebase or deflection coil faults, and any trace of color fringes indicates gun CONVERGENCE faults.

dot matrix a method of displaying letters or digits as a pattern of dots. The method is used mainly in computers both for CATHODE-RAY TUBE displays and for printing on paper.

double-base diode see UNIJUNCTION.

double-beam CRT a CATHODE-RAY TUBE with two independent

electron guns, producing two traces on the screen. The guns are usually connected so the same TIMEBASE is applied to both guns, which also share focus and brightness circuits. The Y-PLATES are independent so the two traces can display different signals. The double-beam tube is particularly useful for displaying phase differences between sine waves and time differences between pulses. Formerly, split-beam guns were made that used a single gun and split the beams after the focus electrode. This provided two-beam operation, but with the disadvantage that one signal was displayed in inverted form. Some oscilloscopes simulate double or multiple beam action by BEAM SWITCHING.

double-pole switch a SWITCH that has two unconnected contact sets that are mechanically but not electrically connected, so two separate circuits can be switched simultaneously.

double-sideband transmission the TRANSMISSION of both of the SIDEBANDS generated by conventional amplitude modulation. Broadcast radio transmissions other than FM use double-sideband transmission with full carrier amplitude.

double superheterodyne receiver or **triple detection receiver** a SUPERHETERODYNE RECEIVER with two conversion stages. The system is often used for receivers that have to cover a wide range of input frequencies, such as a communications receiver. One IF will be high frequency, 10MHz to 30MHZ, and the second IF will be at the conventional 470kHZ. The system enables narrow-band reception, good selectivity, and no risk of FREQUENCY PULLING of the local oscillators.

double-throw switch a switch with two closed (on) positions in addition to the open position. A *single-throw switch* is a switch with one closed (on) position in addition to the open position.

double-tuned circuit a CIRCUIT containing two inductors that have mutual INDUCTANCE and are each part of a tuned circuit.

down-converter a frequency changer circuit whose output is at a lower frequency than its input. The term usually denotes a converter from SHF or MICROWAVE frequencies to normal TV frequencies as, for example, direct reception through a dish from a satellite.

down-counter a digital counting circuit whose output count number decreases for each input pulse.

drain the charge-collecting electrode of a FIELD-EFFECT TRANSIS-TOR.

drift 1. a free movement of a beam of electrons that is not being acted on by an accelerating field. Such electrons will drift with a steady velocity. See DRIFT SPACE, KLYSTRON. 2. an unwanted change in quantities such as the frequency of an oscillator or the voltage output of a supply. Drift often occurs in the first few minutes after switching on and is mainly due to heating of components.

drift space a part of an electron-beam tube that is free of magnetic or electrostatic fields that would act on the moving electrons. There are no accelerating fields in this region, so each electron in a beam will continue with the velocity it had when it entered this space. This allows bunching to occur (see BUNCHER RESONATOR).

driven antenna an ANTENNA that is directly connected to a transmitter or receiver. An example is the dipole of a YAGI ANTENNA.

driver a CIRCUIT that provides power. A *line driver* provides output signal power at an IMPEDANCE that matches a transmission line, so power can be transmitted along the line. A *bus driver* is a DIGITAL CIRCUIT that provides current amplification for the data or address signal outputs of a microprocessor, thus permitting a large number of connections to the bus (see FAN-OUT). A *driver stage* in an audio amplifier is the STAGE that supplies signal to the output stage.

driver of loudspeaker the electromagnetic part of a loudspeaker, consisting of VOICE COIL and magnet.

driver transformer a TRANSFORMER connected as a DRIVER. The term usually denotes a transformer that enables a single-ended transistor stage in an audio amplifier to provide signals to a complementary pair, or to a more complex single-ended push-pull circuit.

driving impedance the IMPEDANCE of a driven CIRCUIT usually a line, or of a power-absorbing electromechanical system. A loudspeaker, for example, has a driving impedance that is composed of its coil resistance and its MOTIONAL IMPEDANCE.

droop or **sag** the reduction of voltage level of a pulse top. A square pulse ought to have a pulse top that is of constant voltage. If this voltage decays toward zero, it is said to droop. The amount of droop is the difference between the ideal voltage level and the true level at the end of the top section. See Fig. 26.

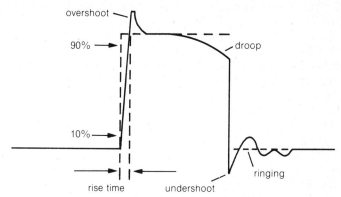

FIG. 26. **Droop.** The droop of a pulse shape. The figure also illustrates other faults, such as rise time, overshoot, undershoot, and ringing.

dropin a piece of nonmagnetic material on a tape. A dropin is caused by contamination and will result in momentary loss of signal when the tape is played.

dropout a piece of tape with missing magnetic coating. Signal cannot be recorded on a dropout section, so this also results in loss of recorded signal.

dropper a resistor that dissipates power by heating and reduces circuit voltage. The term usually denotes a resistor in a circuit fed by a DC source.

dry joint a faulty soldered connection. It occurs when the SOLDER has not been hot enough, has not been applied to the joint for long enough, or when the flux has not burned off completely. In a dry joint, there is a high resistance between each metal and the solder, and the joint may also be a RECTIFYING CONTACT.

DTL *abbreviation for* diode-transistor logic, an obsolete system once used for logic integrated circuits.

D-type flip-flop a form of integrated circuit FLIP-FLOP that has a CLOCK input and a data input. The data input (the D of the name) is normally disabled until the clock pulse arrives. When a clock pulse arrives, data present at the D input are transferred to the device output and held until the next clock pulse.

Some types of D-type flip-flop will LATCH at the leading edge of the clock, some at the trailing edge, some on the top. An *edge-triggered* D-type is a flip-flop that latches on one edge, leading or trailing according to the design.

dual-gate FET a type of signal mixer. The FIELD-EFFECT TRANSISTOR channel is controlled by two gates, so that a signal on either or both gates will affect the channel current. It forms an excellent mixing circuit for high-frequency receivers, because there is virtually no connection between the two gates. The device is used to a large extent in FM and TV receivers.

dual-in-line (DIL) see DIL PACKAGE.

dubbing the mixing of recorded sound SIGNALS to form a single signal. The process is commonly used with multitrack recording methods, in which each instrument has its own microphone and provides a separate recording track. Dubbing enables the sound engineers to maintain good control over the sound of a recorded performance, but it sometimes produces results unlike a live performance.

dummy antenna a network of passive components that acts as a load for a transmitter. The component values are selected so that the network simulates the action of an antenna at the frequency of transmission, but with no radiation. Transmitter power is dissipated as heat, and the resistors on the antenna must be capable of dissipating this power. Use of a dummy antenna permits final adjustments and measurements to be made on a transmitter before connection to a real antenna.

dummy load a power-absorbing circuit that acts like a LOAD and is used for test purposes. A loudspeaker, for example, can be replaced by a resistor of the correct power rating and value, about 8 ohms. This will not allow the effects of MOTIONAL IMPEDANCE to be simulated, but it does allow testing of such features as the power-frequency characteristics.

duplex a two-way simultaneous signaling operation such as a telegraph system in which messages can be sent over the lines in both directions at the same time. Compare HALF-DUPLEX, SIMPLEX.

dust core or **powdered-iron core** a type of magnetic core for an indicator. A dust core is made of powdered magnetic material bound by a resin. The dust is itself an insulator so that EDDY CURRENTS cannot flow, and the main source of losses is HYSTERESIS. Dust cores are used in coils for frequencies in the range 100kHz to 100MHz, but the materials used for high-frequency signals must be selected carefully.

duty cycle a series of repeated operations, such as a SYSTEM that is pulsed on and off (see PULSE).

duty factor or **duty ratio** the ratio of on-time within a duty cycle to total duty-cycle time for a DEVICE.

DVM see DIGITAL VOLTMETER.

dynamic (of a SYSTEM) using under working conditions with changing or varying SIGNALS applied. A dynamic measurement, for example, means a measurement on a circuit that is operating with a signal input, as distinct from a measurement made with no signals, or with power switched off.

dynamic convergence the CONVERGENCE of a set of electron beams when scanned. For a COLOR TELEVISION tube, static convergence means that the three electron beams of the guns strike the correct phosphor stripes at the center of the tube. When the tube is scanned, however, the convergence may not be correct at the edges. This is a dynamic convergence problem that is solved by applying correcting signals, which are derived from the time-base waveforms, to the convergence coils.

dynamic memory see DYNAMIC RAM.

dynamic RAM (DRAM) a type of semiconductor MEMORY in which each BIT is stored as CHARGE on a semiconductor CAPACITOR. The charge leaks away unless regenerated by special REFRESH circuits, but the principle can be used to construct single chips with very large memory storage.

dynamic range the available range of output signal from a device. The lower limit of output is set by the noise level, and the upper limit is set by BOTTOMING or SATURATION. The dynamic range is usually given in decibels.

dynamic resistance 1. the resistance of a parallel resonant circuit at the resonant frequency. **2.** the resistance of any device measured with AC of a stated frequency.

dynamic tracking a method of improving VIDEO CASSETTE RECORDER performance. Conventional video recorders use revolving heads that scan across the moving tape at high speed. Because of mechanical tolerances, there must be a guard band of un-recorded tape between recorded bands.

If the recording and replay head is mounted on a PIEZOELEC-TRIC transducer, it is possible to lay down much more closely spaced tracks, and to follow any track by tiny adjustments of the head position, which are achieved by a servofeedback system.

dynamo a generator that uses a rotating coil (*rotor*) and a magnetic field (*stator*). By convention, a dynamo picks up current from the coils by using a COMMUTATOR and brushes. This gives an output that, for a single coil, is the same as that from a FULL-WAVE REC-TIFIER. If slip rings are used for making connections, the machine is described as an ALTERNATOR.

Today, it is more common to use a magnetic rotor fed with DC and to take the AC from the stator to a rectifier system in order to obtain DC. Such a system is easier to control by electronic methods, using the electronic system to control the current to the rotor.

dynode an ELECTRODE that acts as a multiplier of electrons. When a fast-moving electron strikes a material, it may be reflected or absorbed, or it may cause SECONDARY EMISSION. If the accelerat-ing potential is in the region of 200V to 2,000V, secondary emis-sion is more common. This means that for each electron incident on the surface, more than one electron will be emitted. For some materials, notably elements that are also photoelectric emitters, such as cesium, the number of secondary electrons per primary electron is high, of the order of three to seven.

A dynode consists of a metal mesh or perforated plate that is coated with a material that has a high secondary-emission ratio. When an electron beam strikes this dynode, the beam leaving the dynode is much denser in electrons than the beam striking. In other words, the beam current has been amplified. Because this type of amplification is virtually noiseless, it is used extensively in photoelectric devices such as photomultipliers as a way of obtain-ing great sensitivity.

dynode chain a series chain of resistors with a dynode fed from each junction of two resistors. This chain ensures that the correct relative accelerating voltages are applied to each dynode, typi-cally in a chain of four to eight dynodes.

E

EAROM *acronym for* electrically alterable read-only memory, a form of NONVOLATILE computer memory whose contents can be altered by electrical signals while it is connected in a circuit.

earphone a miniature electrical sound TRANSDUCER. The earphone is designed to be used in contact with the ear, so comparatively good sound quality can be achieved with none of the complications of loudspeaker enclosures, and with much lower power levels.

eccentric line a type of COAXIAL CABLE in which the inner electrode is not central.

E-cell an electrolytic timer that is inexpensive to construct and can be used to indicate service intervals. The principle of the device is based on the electrolytic deposition of a metal spike, with the length of the spike indicating the time for which current has flowed.

echo a reflected wave of sound or electromagnetism. The return beam of a RADAR system is called an echo. TV signals, particularly on UHF, can be reflected by metal objects and cause disturbing effects. For example, if an echo is strong and not too far out of line from the main signal direction, it will cause a second image on the TV receiver screen. This is particularly noticeable if the echo signal is delayed, because this causes the second image to be displaced. An echo image of this type is called a *ghost*. Ghosting is dealt with by using better DIRECTIONAL ANTENNAS or by shifting the antenna position. Ghosting can also be caused by a long and incorrectly terminated transmission line (see TERMINATION).

echo sounding the location of underwater objects by using RADAR methods with sound waves in water.

ECL see EMITTER COUPLED LOGIC.

eddy current an unwanted current flowing within the conducting core of a transformer, causing energy loss. Eddy currents are caused by voltages induced by the varying magnetic field around a transformer. They can be reduced by dividing the transformer core into a number of thin plates, or laminations. See also EDDY CURRENT HEATER.

eddy current heater or **induction heater** a method of heating metal objects without making physical connections. The object is surrounded by or placed near a coil in which a large current of high-frequency signal is flowing. The induced eddy currents will then heat the metal. Metals can be heated in a vacuum in order to remove absorbed gas, or melted by using an eddy current heater. This type of heating often occurs in TRANSFORMERS as an unwanted effect.

edge connector a connector formed as part of a printed circuit board. The metal at the edge of the board is shaped into a set of strips, and each strip is connected to some part of the circuit. Connections are made by a plug that grips the board and uses metal springs to make contact with each strip.

edge effect the existence of a nonuniform, rapidly changing electric field at the edge of a capacitor plate.

effective resistance the RESISTANCE to AC measured by the dissipated power. A figure of resistance is obtained by measuring power and dividing by the square of current, or by dividing the square of voltage by power. This may not correspond to the DC resistance value, because it includes a resistance equivalent for power lost because of eddy currents, hysteresis, and the like.

effective value 1. a value of some quantity measured under operating conditions. 2. the ROOT MEAN SQUARE value of an alternating quantity.

efficiency the percentage conversion of energy. The efficiency of the power amplifier stage of a radio transmitter is the output power as a percentage of the DC supply power (*input power*) to the stage.

EHF *abbreviation for* extremely high frequency millimetric waves in the wavelength range 1mm to 10mm.

EHV *abbreviation for* extra high voltage, for example, high-voltage supplies for CATHODE-RAY TUBES, CAMERA TUBES, DYNODE

CHAINS and radio TRANSMITTERS. Any voltage higher than 1,000V is in the EHV range. See also HIGH TENSION.

elastoresistance the change of electrical resistance with stress. Stretching a film of a conductor will lengthen the specimen and also cause its cross section to decrease. Both effects will also cause the resistance to rise. Elastoresistance forms the basis of stress measurement in STRAIN GAUGES.

E layer or **Heaviside layer** or **Kennelly-Heaviside layer** the layer in the IONOSPHERE that lies approximately between 90 and 150 km above Earth's surface. It reflects the medium-frequency range and was one of the first of the ionospheric layers to be identified. Oliver Heaviside (1850—1925) predicted the possibility of the ionosphere, and Arthur Kennelly (1861—1939) proved the existence of the layers and measured their heights by using RADAR techniques.

electret a permanently charged insulating material. Electrets are made by allowing molten plastics to solidify in a large electric field. They have been used in constructing capacitor microphones and phonograph pick-ups.

electric axis the direction in which the maximum potential difference is developed when a PIEZOELECTRIC crystal is stressed.

electric eye see PHOTOCELL.

electric field a field of force surrounding a charged particle within which another charged particle experiences a force. The intensity of the electric field at a point is equal to the force-per-unit charge placed at that point. The symbol for electric field is **E** where

$$\mathbf{E} = \frac{\text{Force}}{\text{Charge}} = \frac{\mathbf{F}}{Q}.$$

Q = charge and is measured in coulombs. $\mathbf{F}$ = force and is measured in newtons per coulomb.

electric potential or **voltage** a measurement of the electrical equivalent of height. Symbol: V; unit: VOLT. The potential at a point is usually measured as POTENTIAL DIFFERENCE from GROUND POTENTIAL and is defined as the amount of work done per unit CHARGE when a charge is removed from that point to infinity. Potential and potential differences are easily measurable.

electroacoustics the study of devices that work with electric and sound signals, including MICROPHONES, LOUDSPEAKERS, and all forms of sound TRANSDUCERS including ULTRASONIC transducers.

electrocardiograph (ECG) an instrument for displaying and recording heart currents. The operation of the heart, like the operation of any other muscle, is controlled by small electrical signals that can be picked up on probes and displayed on a CATHODE-RAY TUBE or on a chart recorder. An electrocardiograph normally uses BALANCED AMPLIFIERS because of the very small voltages that are being picked up. Grounding is important, because any fault current transmitted to the probe electrodes can be fatal.

electrode a conductor in an electronic device that is at some ELECTRIC POTENTIAL and can collect or emit charged particles.

electrodynamic (of an electrical device) operated by the mechanical forces arising from magnetic fields due to wires carrying CURRENTS particularly in coil form.

electroencephalograph an instrument that detects and displays the voltage waveforms of the brain. For general techniques, see ELECTROCARDIOGRAPH.

electroluminescence the giving out of light from a material when an ELECTRIC FIELD is applied. Materials that are electroluminescent will give out light when subjected to an alternating electric field, using a construction like a capacitor with one transparent plate. The effect has fallen out of widespread use because of the short operating life of the material.

electrolysis the effect of the flow of IONS in a molten or dissolved solid. By immersing a pair of conducting plates in the material, one positive (*anode*), the other negative (*cathode*), the ions can be collected. The combination of ion collection with chemical changes can be a valuable source of pure materials, including some elements that cannot be prepared in any other way, for example, aluminum. See also CATION, ANION.

electrolyte a liquid that conducts because of the presence of IONS. Conducting liquids include molten salts and water solutions of acids, salts, and alkalis. Some ionic materials such as liquid ammonia also form ionic solutions.

electrolytic capacitor a CAPACITOR that provides large capacitance values in a small volume. The simplest electrolytic capacitor consists of aluminum sheets sandwiching a piece of blotting paper that has been soaked in aluminum perborate. Applying a voltage to this arrangement causes a thin film of insulating oxide to form on one sheet, and it is the large capacitance across this film that is utilized. The capacitance value can be increased by increasing the area of one plate by perforating it and folding it. The capacitor is sealed to avoid evaporation of the electrolyte liquid.

Though high values of capacitance can be achieved, there is always a comparatively large leakage current. The capacitor is POLARIZED (1), which means that the voltage must always be applied to it with the same polarity that was originally used to form the insulating layer. Failure to observe this can lead to explosive failure. The voltage that can be permitted across the capacitor is also low, and few electrolytic capacitors can be used at DC voltage levels of over 500V. The capacitors are also temperature sensitive, and the voltage rating must be drastically reduced when ambient temperatures are high. Many specifications for military equipment, or equipment to be used under a wide range of environmental conditions, do not permit use of electrolytic capacitors.

electrolytic polishing a method of obtaining a high polish by electrolytically filling up any irregularities on the surface of a metal. The metal forms the ANODE in an electrolytic cell.

electromagnet a magnet that has a high FLUX DENSITY value only when current flows through coils. Any coil of wire will act as an electromagnet, but the effect is enhanced when a core of FERRO-MAGNETIC material such as soft iron, a SOFT MAGNETIC MATERIAL is used.

electromagnetic deflection the deflection by electromagnets of an ELECTRON BEAM such as a CATHODE-RAY TUBE beam. The neck of the tube is surrounded by a set of coils. When a current of SAWTOOTH WAVEFORM flows in the coils, the beam is deflected to form a linear TIMEBASE. Two pairs of coils are used to form a TV RASTER scan.

For RADAR use, a radial scan is used, and this is usually done using three coils at 120 degree angles. The phase of the sawtooth current in the three coils is controlled so as to alter the position of the end of each scan.

electromagnetic focusing the focusing of an ELECTRON BEAM by electromagnets. This technique is no longer common, having been replaced by electrostatic focusing in modern TV receiver tubes. Electromagnetic focusing was used on early designs and for some radar tubes.

electromagnetic induction the production of a POTENTIAL DIFFERENCE by changing MAGNETIC FLUX. A potential difference will be produced between two points if there is a change in magnetic flux at right angles to a line drawn between the points. If the two points are joined by a conductor, the potential difference can be measured and used.

This is the basis of generators, alternators, and dynamos. The size of potential difference is equal to the rate of change of flux (Faraday's law), and its polarity is such that if current flowed, it would set up a flux change in the reverse direction (Lenz's law). If no conductor is present, the changing magnetic flux causes an electric field.

electromagnetic lens a lens for ELECTRON BEAMS produced by using coils to form a magnetic field that will focus the electron beams. An electromagnetic lens is easier to work with than an ELECTROSTATIC LENS when high energy electron beams are used, as in electron microscopes and some electron beam tubes.

electromagnetic radiation the radiation of energy in the form of varying electric and magnetic fields—static fields do not cause wave radiation. Radiation is a consequence of the laws of electrostatics and of electromagnetism, which are gathered together in MAXWELL'S EQUATIONS. The fundamental cause of electromagnetic radiation is the acceleration of charged particles, so any alternating current is capable of causing radiation. In practice, radiation is achieved efficiently only for frequencies of the order of 100kHz upward. Light is a form of electromagnetic radiation. See ELECTROMAGNETIC SPECTRUM.

electromagnetic spectrum the range of known and possible electromagnetic waves. This range extends from very long waves, low frequency, at 100kHz or even less, up to very short wavelengths (high frequencies) such as x-rays and gamma radiation. The spectrum includes all known radio waves, including many that at present cannot be easily generated or detected—infrared, light, ultraviolet, and all the wave frequencies that arrive from outer space. See ELECTROMAGNETIC RADIATION.

electromagnetism MAGNETISM caused by electric CURRENTS. The term is commonly used to denote magnetism caused by passage of electric currents through coils of wire, but magnetism produced by any moving charge, as in a radiated wave, is electromagnetism, as is permanent magnetism.

electrometer a form of VOLTMETER with high input resistance. The term was once used to denote electrostatic voltmeters, but is now applied to FET and vacuum tube voltmeters that can be used to measure potential differences across very high resistances.

electromotive force (emf) the voltage of a source of electrical energy. The term POTENTIAL DIFFERENCE is usually reserved for voltage across a device that is dissipating electrical energy.

electron a stable, negatively charged elementary particle orbiting the positive nucleus of an atom. The number of electrons in a neutral atom equals the number of positive PROTONS in the central nucleus, which is equivalent to the atomic number of the element.

The charge of an electron is 1.6022×10^{-19} coulomb. This charge is of equal magnitude but opposite sign to that of a proton. Electrons have a very small rest mass (9.1096×10^{-31} kilogram, 1836 times smaller than that of a proton), and can easily be detached from atoms in a vacuum and accelerated to high speeds. The movement of electrons is what we call electric CURRENT. Solid conductors are materials in which some electrons (the conduction electrons) are loosely attached to atoms and will move readily through the solid material.

electron beam or (formerly) **beta ray** a large number of electrons flowing in the same direction in a confined path. An electron beam is achieved by using ELECTROSTATIC FIELDS to accelerate electrons away from a source, and either electrostatic or magnetic fields to form the scattering electrons into a beam. See also ELECTRON LENS.

electron gun an arrangement of an electron-emitting CATHODE and ELECTRON LENSES that forms an ELECTRON BEAM. The device is used in CATHODE-RAY TUBES.

electron-hole pair the charged particles produced by heating or

DOPING a SEMICONDUCTOR. The pure (intrinsic) state of semiconductors is almost free of charged particles that can move. By heating or by adding doping impurities, electrons can be released, leaving gaps in the structure of the materials (HOLES). These holes behave like charged particles, with positive charge and an apparent mass approximately equal to the mass of the electron. Since there must be a hole for each free electron, these particles always exist in pairs in a semiconductor.

electronic device any device that achieves its effect by directly controlling the movement of ELECTRONS or HOLES. The electrons may be moving in a vacuum, a solid, or a gas. Their movement can be accompanied by the movement of holes or ions.

electronic memory a form of MEMORY in which writing, storage, and reading are electronic. The storage action can use CHARGE storage (electrostatic memory), the switching of an electron flow between devices (flip-flop static memory), or magnetization of a magnetic material (as in core storage).

electronics 1. the science and technology concerned with the development, behavior, and application of devices that use moving electrons, ions, or holes. 2. the circuits and devices of a piece of electronic equipment.

electronic switch a SWITCH with no moving parts. Switching is carried out by acting on the electrons or other CARRIERS with electric or mnagnetic FIELDS. This makes switching fast, and there are no problems of contact or wear.

electronic tuning the changing of the FREQUENCY of a tuned circuit without mechanical intervention. Electronic tuning by means of VARACTOR diodes is extensively used for FM radio and TV receivers.

electron lens a set of ELECTRODES or coils that can focus an ELECTRON BEAM. The electrons from a CATHODE are accelerated through a hole at the end of a grid cylinder that is at a negative potential. The shape of the fields at this aperture causes the electron paths to cross over, with a point of minimum diameter. This is the start of a beam, and the diverging paths of the electrons from this point are then controlled either by using ELECTROSTATIC FIELDS or by using MAGNETIC FIELDS.

electron microscope an instrument for viewing images of very small objects. The microscope is based on the principle that a fast-moving particle will exhibit some of the properties of a wave, so an ELECTRON BEAM behaves like a shortwavelength light beam. The electron beam can be arranged to pass through the specimen or to be reflected from it. Greater resolution is achieved by this means than is possible optically, because the wavelength of an electron beam is much shorter than that of light.

The *transmission electron microscope (TEM)* enables electrons to pass through the specimen, and the resulting image is viewed on a fluorescent screen. Considerable experience is needed to interpret the images correctly.

The *scanning electron microscope (SEM)* uses TELEVISION techniques, scanning the electron beam across the object and picking up the scattered, reflected, and secondary electrons from the surface. These returned electrons are then used to form a TV-type of output signal. Microscopes of this type yield an image that is much easier to interpret and looks like the image seen through an optical microscope.

electron multiplier a set of DYNODES arranged to multiply the number of electrons in a beam at each dynode stage. The principle is used particularly in photoelectric devices as a way of noiseless amplification.

electron optics the study of the control of ELECTRON BEAMS. Electron optics includes the production, focusing, deflection, and use of electron beams.

electron-volt the amount of energy acquired by an electron accelerated by one volt. Symbol: eV. A useful unit for measuring small energy changes, equal to 1.6×10^{-19} joules. See also WORK FUNCTION.

electroplating the deposition of metals from a solution onto a conductor. The object to be coated is suspended from wires in a bath of an acid or alkaline solution that contains salt of the metal to be used for coating. A bar or sheet of the coating metal (*anode*) is also immersed in the solution, but not touching the object to be coated. The two are then connected to a supply, the anode to the positive pole, using a low voltage, typically four to eight volts. By movement of ions and chemical actions, metal is dissolved from the anode and deposited onto the object to be coated. Electroplating is used extensively for coating metal objects with gold, silver, chromium, nickel, and other metals.

electrostatic of, concerned with, producing, or caused by electric CHARGES and ELECTRIC FIELDS at rest.

electrostatic deflection the deflection of an ELECTRON BEAM that occurs when electric fields are applied between metal plates. The beam is deflected in the direction of the more positive plate. See DEFLECTION SENSITIVITY.

electrostatic field the ELECTRIC FIELD that acts with force on an electron or ion. An electrostatic field exists around any charged particle, and around any conductor that is connected to a potential.

electrostatic focusing the focusing of an ELECTRON BEAM by passing it through shaped metal electrodes at different potentials. The aim is to create a SADDLE-SHAPED FIELD that will converge a diverging beam so it focuses on some target, such as a PHOSPHOR screen.

electrostatic induction the production of electric CHARGE on an object, whether a conductor or nonconductor, because of presence of another charge. Electrostatic induction is caused by separation of electrons from atoms under the influence of an electrostatic field from another charge or potential.

electrostatic lens an assembly of metal electrodes at different ELECTRIC POTENTIALS. The fields set up will cause an electron beam to converge or diverge, just as an arrangement of glass lenses will affect a beam of light. Compare ELECTROMAGNETIC LENS; see also SADDLE-SHAPED FIELD.

electrostatic loudspeaker a loudspeaker whose operating principle is the deflection of a charged membrane by electrostatic forces. The technique is used for high quality sound reproduction.

electrostatic precipitator a method of removing dust in, for example, an air-conditioning system. Air is forced through a metal mesh that is maintained at a DC potential of several kV. Dust particles, which are usually charged, are attracted to the mesh, which can be cleaned at intervals by switching off the potential.

electrostatic separator a method of separating powders of insulating materials that are falling or moving along a belt. The principle is to apply a strong ELECTROSTATIC FIELD so that materials with different permittivity values experience different amounts of force and take different paths.

electrostriction the movement of the atoms of a material caused by ELECTROSTATIC FIELDS. This means the materials will expand or shrink by a very small amount when the field is applied or removed.

element a material made up of identical atoms.

emf or **EMF** see ELECTROMOTIVE FORCE.

emission the release of electrons from a material. The term usually denotes the release of electrons into a vacuum from a metal under the influence of heat (THERMIONIC EMISSION), light (PHOTOEMISSION) or electric fields (FIELD EMISSION).

For emission to be achieved, each electron must acquire an amount of energy greater than the binding energy that holds it in place. Electrons in semiconductors are displaced, not emitted from the material.

emitter 1. the electrode in a semiconductor device from which electrons or holes pass to another layer of the material. 2. another term for CATHODE.

emitter-coupled logic (ECL) or **current-mode logic (CML)** a fast-acting type of logic circuit used in computing. The ECL circuit employs TRANSISTORS that are connected at the emitters to a common LOAD. The load is such that the current through it is constant. Small changes of bias voltage at the bases of the transistors will cause current to be switched from one transistor to the other. The voltage needed for switching can be very small, and the amount of current switched can be comparatively large, allowing STRAY CAPACITANCES to be charged or discharged rapidly. An ECL device can usually be switched at rates of up to 200MHz. See Fig. 27.

emitter follower or **common-collector connection** or **grounded-collector connection** a circuit with high input IMPEDANCE and low output impedance. The load is placed in the emitter circuit, and the input signal is applied to the base. Because the base-emitter voltage of a transistor varies only slightly when a small signal is applied, the emitter voltage follows the base voltage very closely, hence the name emitter follower.

The circuit is used extensively as a BUFFER and a DRIVER for

ENABLING

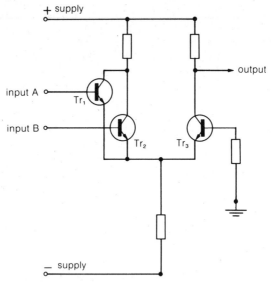

FIG. 27. **Emitter-coupled logic.** With inputs A and B at zero, Tr_3 conducts. When either A or B or both are high, transistor Tr_1 or Tr_2 or both will draw current at the expense of Tr_3. This will cause the voltage at the collector of Tr_3 to rise. This is an elementary OR-gate.

loads with high stray capacitance. The use of emitter followers as buffers has largely been superseded by the introduction of OPERA-TIONAL AMPLIFIER followers. See Fig. 28.

enabling (of a circuit or device) allowing to act normally. For example, a circuit can be disabled by removing or reversing BIAS so that an input signal is ineffective. Digital devices are often provided with an enable input to activate/deactivate them as required.

encoder a device that produces an output in electronic form corresponding to some measurement. The output is usually in digital signal form. For example, a shaft-position encoder is a form of TRANSDUCER that produces digital outputs corresponding to angular positions of a rotating shaft.

endaround shift see ROTATION.

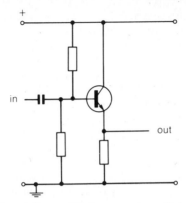

FIG. 28. **Emitter follower.** This is a current amplifier, with a voltage gain of less than unity. In the figure, base-bias resistors are shown, but a more common method is to couple the base directly to the previous stage.

endfire array an ANTENNA array in which the aiming direction is along the line of antennas rather than at right angles to the line.

energy band a permitted range of energy values for particles in a solid. An energy value within this range is called the *band energy.* In single atoms, each electron has a definite amount of energy, called its energy level, and no intermediate levels of energy are found. When atoms are closely spaced, as in solids, these permitted energy levels merge into bands, so electrons can take any of a range of energy values.

The arrangement of energy bands in a material, and whether bands are full, empty, or partly filled determine whether a material is an insulator, semiconductor, or conductor. A full band (insulator) is one in which one electron exists with each possible level of energy. An empty band (conductor) is one in which there are no electrons with any of the permitted levels of energy. A partly filled band (semiconductor) is explained by its name.

enhancement mode a mode of operation of some types of FIELD-EFFECT TRANSISTORS. A FET that needs enhancement mode biasing will pass little or no current in the absence of a positive BIAS voltage, and a bias voltage has to be applied in order to pass working current. Compare DEPLETION.

envelope the outline of the amplitude peaks of a modulated radio wave. This envelope shape should be the same as the shape of the wave being used to modulate the radio frequency.

epitaxial transistor a (PLANAR) transistor that has been formed by EPITAXY. This means that a layer of lightly doped material has been deposited on a heavily doped main collector region. This extra layer affords better collector-base BREAKDOWN characteristics, while the heavily doped region permits a low-resistance collector connection and good THERMAL CONDUCTIVITY.

epitaxy the growth of other layers on the surface of a SEMICONDUCTOR crystal, usually from a gas. The layers will be of doped or INTRINSIC material, and the importance of the process is that the added layers maintain the same crystal structure as the underlying crystal. As a result, the entire structure behaves like one crystal. Epitaxy has been the main process for creating transistors and integrated circuits for many years. However, it is being superseded by ION IMPLANTATION for some specialized purposes.

EPLD *abbreviation for* erasable programmable logic device, an assembly of identical gate and inverter circuits whose connections can be established by programming and can be reprogrammed. An EPLD is a form of UNCOMMITTED LOGIC ARRAY.

epoxy resin a plastic resin with exceptionally good insulating characteristics. The value of epoxy resins is that many are particularly easy to use in coating or molding applications. Most of these resins can be obtained as two-pack types, consisting of a viscous resin and a liquid setting agent. The two are mixed just before use, and the setting time depends on the mixture ratios and on temperature.

EPROM *acronym for* erasable programmable read-only memory, a semiconductor memory integrated circuit, or chip, for computer use in which data are held semipermanently. An EPROM can only be erased by shining an intense ultraviolet light through a transparent quartz window in the chip. Once it has been erased, data can be stored in an EPROM and will be retained even when no power is supplied to the chip.

equalization a form of controlled DISTORTION. The best example is the treatment of audio signals for disk recording. The bass signals are reduced in amplitude, because the amplitude of the groove wave would otherwise be too great. Similarly, the amplitudes of high-frequency signals are boosted, because the disk noise is mainly in this region. After transcribing, that is, after converting back into electrical signals, the bass amplitude has to be boosted and the treble attenuated in order to recover the correct signal balance.

equalizer any circuit that carries out EQUALIZATION. The term is used also to denote circuits that will selectively boost or attenuate parts of the audio spectrum. Equalizers are used in professional applications to overcome studio problems of boom or deadness. The device is sometimes called a *graphics equalizer* particularly when the control panel shows the effect of the controls on a frequency response graph.

equivalent circuit an arrangement of simple electrical COMPONENTS that is electrically equivalent to a complex CIRCUIT. The equivalent circuit is used for purposes of calculation and makes use of simple real components, such as resistors, inductors, and capacitors, along with idealized components, such as signal voltage or current generators.

By representing a complex device such as a transistor voltage amplifier with an equivalent circuit that may, for example, contain a voltage generator and a few resistors, it is possible to calculate gain and other features closely enough for practical purposes. See also NORTON'S THEOREM, THEVENIN'S THEOREM.

equivalent resistance a single resistance value that has the same effect as a number of other RESISTORS or power-dissipating components. A circuit may, for example, contain RESISTANCE and INDUCTANCE. Its equivalent resistance value will allow calculation of total power dissipated in all the parts of the circuit, including the resistance of the inductor and the impact of any effects such as motional impedances. Use of an equivalent resistance is particularly suited to the treatment of the inputs and outputs of ACTIVE COMPONENTS.

erase remove data from a memory store, usually semiconductor memory, but also magnetic media such as tapes or disks.

erase head a tape recorder head used to erase tapes or disks immediately prior to recording. The erase head uses a comparatively large gap and is fed by a high-amplitude, high-frequency signal. The effect is to take each part of the tape through a large number of HYSTERESIS cycles, gradually reducing the peak amplitude so that the magnetic particles are left unmagnetized.

etching the dissolving of material by a chemical process, particularly with acids. Etching is used to produce printed circuit boards, using solutions of materials such as ferric chloride. Etching is also used in the production of transistors and integrated circuits.

excitation the addition of energy to a system, for example, signals that form an input for a system such as an amplifier or oscillator.

exclusive-OR (XOR) gate a type of LOGIC GATE giving a logic comparison. The OR GATE produces an output of 1 if any or all of its inputs are at logic 1. The XOR gate excludes the case of more than one level 1 input, so the XOR gate output is 1 if only one input is at level 1. See Fig. 29.

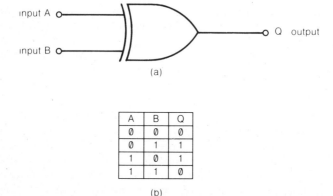

(a)

A	B	Q
0	0	0
0	1	1
1	0	1
1	1	0

(b)

FIG. 29. **Exclusive-OR gate.** The two-input exclusive-OR gate symbol (a) and truth table (b). The logic of the gate is that the output is 1 only for one single input equal to 1.

expanded sweep a TIMEBASE for a CATHODE-RAY OSCILLOSCOPE in which the timebase amplitude can be greatly increased. This allows the effect of time magnification, so a small part of a wave-form can be examined in detail without switching to another time-base speed and resynchronizing.

expander a circuit that increases the dynamic range of a signal. The action is achieved by a NONLINEAR stage whose gain increases as the signal amplitude increases. See also COMPANDER.

F

facsimile (fax) the electronic TRANSMISSION of text, diagrams, and the like. This was formerly done by electromechanical methods, but now is achieved by computing techniques. The material to be transmitted is scanned, and each unit area shape generates a BINARY CODE, and it is this code that is transmitted. At the receiver, each code operates a set of pins in a DOT-MA-TRIX printer, producing the same pattern that existed in the material transmitted.

fader a form of ganged ATTENUATOR normally using a single control. A fader allows one signal to be reduced to zero amplitude smoothly while another signal is increased from zero to normal amplitude.

fading the reduction of CARRIER (1) signal strength at a receiver. Fading is usually caused by interference between waves that arrive at the receiver by different routes. At least one such path must be caused by reflection of a wave by the IONOSPHERE and the reflecting height of the ionosphere is continually shifting. Because of this, waves at the receiver produce a resultant that can be of low or high amplitude according to the phase angle between the arriving waves. See Fig. 30. Fading is counteracted by DIVERSITY reception and by using AUTOMATIC GAIN CONTROL.

failure rate a measure of the unreliability of a component or system. The failure rate is measured in terms of the number of failures per thousand hours of operation, or thousands of on/off cycles, as appropriate. For some systems using integrated circuits, the failure rate may be immeasurably small (for example, no system may have failed during testing), so failure rates then have to be computed on the basis of ACCELERATED LIFE TESTS under unnatural conditions. Compare RELIABILITY; see also MEAN LIFE.

fall time the time taken for the amplitude of a pulse to reduce to

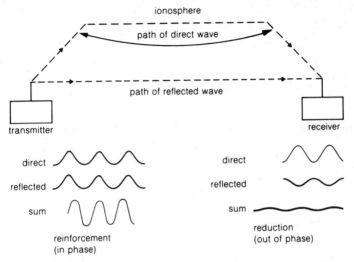

FIG. 30. **Fading.** Fading of a signal is due to the phase differences between the direct wave and its reflection from the ionosphere, or between one reflection and another.

a predefined level, usually measured as the time from 90% of peak amplitude to 10% of peak. Compare RISE TIME.

fan-in the maximum number of inputs that can be connected to a LOGIC CIRCUIT. Compare FAN-OUT.

fan-out the maximum number of LOGIC CIRCUIT inputs of the same family that the output of a device can drive. This quantity is decided by the current available at the output and by the capacitance at each input. Fan-out can be increased by use of BUFFERS. Compare FAN-IN.

farad see CAPACITANCE.

Faraday cage a metal box or mesh screen that is grounded to prevent electrostatic INTERFERENCE between circuits inside the cage and circuits outside. Large Faraday cages are used in testing transmitters or, conversely, to test equipment that might be affected by interference.

fast-recovery diode a diode in which normal forward-biased operation resumes rapidly after a period of reversal. Ordinary diodes

113

suffer from CARRIER-STORAGE effects, which make shutoff comparatively slow. For microwave signals, it is necessary to use SCHOTTKY DIODES or diodes fabricated of compound semiconductors such as GALLIUM ARSENIDE.

fault current a CURRENT usually large, that flows because of a circuit fault, usually a failure of insulation.

FCC *abbreviation for* Federal Communications Commission, the regulating body for radio broadcasting in the US.

feedback the return as part of the INPUT to an AMPLIFIER of a signal that is obtained from the OUTPUT. If the signal fed back is in phase, the amplifier is supplying part of its own input. This makes the effective GAIN much higher and will cause oscillation if the gain of the amplifier is greater than the loss of the network that connects the output with the input.

If the signal fed back is in antiphase, the feedback is negative. This lowers gain, but also reduces noise and distortion caused by the amplifier. An amplifier that has NEGATIVE FEEDBACK is also stabilized against changes caused by alterations in the characteristics of components, such as transistors and resistors. When the amount of negative feedback is large, the only components that noticeably affect the gain are the resistors or other components in the feedback loop itself. The use of negative feedback is an important feature of the design of LINEAR AMPLIFIERS. See also POSITIVE FEEDBACK.

feedback control loop a feedback LOOP used in a control system. The control system may contain mechanical, thermal, hydraulic, or optical components. If it does, these components are included as part of the feedback loop.

feeder a line that carries radio power from a transmitter to an ANTENNA system.

feedthrough a form of connection that passes through a material. The device is often used to describe a type of component, for example, a feedthrough CAPACITOR. A feedthrough capacitor consists of a pin that is surrounded by a coaxial capacitor, creating a comparatively large capacitance between the pin and the body of the capacitor. The device is used to decouple connections that pass out of or into a circuit through a metal panel.

femto- *combining form* denoting a submultiple of the base word equal to 10^{-15} (one quadrillionth).

ferrite a brittle crystalline material made from oxides of iron and other metals, used as an electrically insulating magnetic material.

ferromagnetic (of a material) exhibiting the same type of magnetic behavior as iron, a characteristic of magnetic materials that have high PERMEABILITY. See also HYSTERESIS LOOP.

FET see FIELD-EFFECT TRANSISTOR.

fiber optics the use of highly transparent glass fibers to carry light signals. The light beams are prevented from escaping from the material by total internal reflection. Because a light beam is an electromagnetic wave with a frequency in the range of 10^{14} to 10^{15}Hz, huge amounts of information can be carried on a single beam.

field 1. a part of space in which a force effect acts. In an ELECTRIC FIELD a force can be detected on any charge. In a MAGNETIC FIELD a force will be detected on any magnetic material, or on a wire that carries current. **2.** (in TV) a half-frame of a picture (see RASTER).

field coil a coil used to generate a magnetic field by passing current through the coil.

field control the adjustment of MAGNETIC FIELD in an electric motor or AC generator. Field control is used to adjust motor speed or AC generator output voltage.

field-effect transistor (FET) a form of transistor in which current flow through a CHANNEL (2) is controlled by means of a GATE (1). Contacts are made on the two end faces of a bar of silicon, which is called the *channel* (the *n-channel* if doped with N-TYPE impurities, and the *p-channel* if doped with P-TYPE impurities). The contacts are called the *source* and the *drain* respectively. When the channel is of the n-type, a p-type region will be diffused into the center of the silicon bar, as shown in Fig. 31. If a voltage is applied between the source and the drain, as shown, electrons will be injected into the channel at the source and be collected at the drain.

Around the p-n junction formed between the gate and the channel, a DEPLETION LAYER will form. The presence of this depletion layer narrows the effective width of the channel as far as current carriers (electrons in this example) are concerned. If the gate is made negative with respect to the source, the width of the

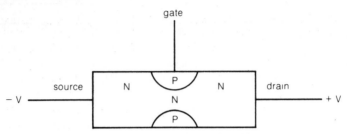

FIG. 31. **Field-effect transistor.** See this entry.

depletion layer increases, reducing the drain-source current. Thus, voltage on the gate controls the current through the device.

If a FET with a p-channel is used, the arguments above still apply, but all voltages must be reversed.

The device described above is known as a *junction FET (JFET)*. It should be noted that the gate of an n-channel JFET cannot be made positive, or conduction will occur from the gate to the channel. A JFET can therefore only be operated in DEPLETION MODE.

If a thin layer of silicon oxide is interposed between the gate and the channel, the gate may be made positive, for n-channel devices, with respect to the channel without gate conduction occurring. This has the effect of widening the gate and reducing the channel resistance. This type of FET is known as an *insulated-gate FET (IGFET)* or *metal-oxide semiconductor FET (MOSFET)* after the type of construction used. MOSFETs can be used in either depletion or ENHANCEMENT MODE.

field emission the EMISSION of electrons in a large electric field. Any sharp point at a high potential can cause field emission, and this can be a cause of failure or flashover, particularly in large vacuum tubes.

field frequency the rate of repetition of the complete screen scan in a TV system. See RASTER.

field magnet a permanent magnet that is used to obtain a MAGNETIC FIELD. The field can be used in a small DC motor or for deflecting an electron beam. An electromagnet is more controllable, so a FIELD COIL is often used.

field-strength meter an instrument used to measure received signal strengths. Field strength is also called *intensity of field.* The meter typically uses tuning circuits, followed by a rectifier to obtain a DC voltage proportional to the amplitude of the carrier. The meter is used in conjunction with an antenna to plot the strength of carrier signal at various places remote from a transmitter. This enables maps of field strengths to be drawn up to assess the effective reception area of a transmitter.

filament a thin metal wire heated by passing CURRENT through it, used in vacuum tubes to provide cathode heating or electron EMISSION. For indirect cathode heating, molybdenum wire is used. For direct emission, tungsten is the most common filament material, though platinum is used in some x-ray tubes and x-ray rectifiers.

filament display an ALPHANUMERIC display that employs tiny metal filaments set into a figure eight pattern. This enables the normal type of SEVEN-SEGMENT DISPLAY decoder to drive the filaments and provide a display brighter than can be achieved with LEDs, albeit with an increased power requirement.

film a thin coating of any material.

film resistor a RESISTOR manufactured by coating a film of metal or carbon onto a ceramic cylinder. A spiral is then cut into the film to adjust the resistor to its nominal value. Film resistors have replaced most of the older resistor types for use in electronics.

filter a NETWORK that can select signals of a required frequency range. Any filter will have a PASSBAND meaning the range of frequencies that pass through it unattenuated, and a STOP BAND of frequencies that are greatly attenuated. The main types are LOW-PASS FILTERS, HIGH-PASS FILTERS, BAND-PASS FILTERS, and BAND-STOP FILTERS. Filters may contain only PASSIVE COMPONENTS such as resistors, capacitors, and inductors or use ACTIVE COMPONENTS such as transistors and integrated circuits. Modern active filters are easier to design than passive filters, although computer-aided design systems make either type reasonably easy to deal with.

Compare ALL-PASS NETWORK.

final anode the ANODE in a CATHODE-RAY TUBE design that collects most of the emitted electrons. The final anode will be a conducting layer either on the screen (see ALUMINIZED SCREEN) or immediately surrounding the screen.

first-in/first-out (FIFO) buffer or **register** a data storage device for computers in which the first data item placed into the buffer or register is the first to be retrieved.

first-in/last-out (FILO) buffer or **register** a data storage device for computers in which the first data item placed into the buffer or register is the last to be retrieved.

fixed return a wave echo from a fixed object such as a building or mountain. The problem posed by fixed returns is that they can make it difficult to detect moving objects. See also RADAR, CANCELLATION CIRCUIT.

flash apply a voltage PULSE to a COMPONENT. Flashing can be used to burn away tiny sharp points caused by intense field emission. It is often used in the burn-in of vacuum tubes.

flashover a temporary breakdown of insulation, allowing a spark or discharge.

flat line a power TRANSMISSION LINE that is perfectly matched to its load, with no trace of STANDING WAVES.

flatpack a thin package that contains an INTEGRATED CIRCUIT. The IC is connected to leads that emerge from the edges of the pack. Compare DIL PACKAGE.

flat tuning a tuning position at which a variation of the tuning control produces little change of signal strength at the DEMODULATOR.

F layer or **Appleton layer** the main reflecting layer in the IONOSPHERE extending from about 150km to 1,000km above Earth's surface. The layer is strongly ionized and will reflect waves with wavelengths of between 6mm and 20m.

Fleming's rules a method of remembering the directional relationship between current, magnetic field and force. See Fig. 32.

flicker the small variations in brightness that are obvious to the eye. Because the eye stores an image for about a tenth of a second, small variations in brightness are not perceived if they occur faster than about 16 to 24 times per second. For bright pictures, however, a much higher repetition rate must be used if the eye is not to notice flicker, hence the use of a field rate of 30 fields per second for TV.

flicker effect the irregular emission of current carriers, causing FLICKER NOISE.

flicker noise a type of noise signal caused by tiny variations in

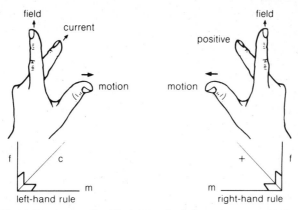

field
current
positive
motion
motion
f   c   m
left-hand rule
m   +   f
right-hand rule

FIG. 32. **Fleming's rules.** The left-hand rule is used to remember the direction of the force on a conductor bearing current in a field of magnetic flux. The right-hand rule indicates which end of a conductor will become positive when the conductor is moved in a field of magnetic flux.

CURRENT. This type of noise exists because current carriers, ELECTRONS or HOLES, are not emitted at a perfectly uniform rate, either into a vacuum or into semiconductor material.

flip chip an INTEGRATED CIRCUIT chip connected directly to a circuit. The flip chip uses thick bonding pads that can be welded to circuit connections when the chip is inverted.

flip-flop a bistable device (see BISTABLE CIRCUIT) used in digital circuits. The output of a flip-flop changes state, from 0 to 1 and vice versa, every time an input pulse is applied. Some flip-flops react to the rising or falling edge of a pulse (*edge-clocked devices*), and some to the height of the pulse (*level-clocked devices*). See also D-TYPE FLIP-FLOP, J-K FLIP-FLOP, MASTER-SLAVE FLIP-FLOP, R-S FLIP-FLOP.

floating (of a device or connection) the state of being unconnected to any voltage or signal.

flood gun an ELECTRON GUN that produces a wide parallel beam of electrons. This is used in a direct-view STORAGE TUBE.

floppy disk a thin, flexible plastic DISK coated with magnetic material and used as a method of data storage for small computers.

fluorescence the emission of light without heating. A fluorescent substance emits light when it is struck by electrons or by high-energy radiation such as ultraviolet radiation.

fluorescent screen see FLUORESCENCE, PHOSPHOR.

flutter a rapid variation in the pitch of a note that is intended to be constant. Flutter usually indicates a fault in a recording or playback system that may be caused by poor speed control of a tape drive or turntable.

Compare WOW.

flux 1. a quantity that represents the effect of a FIELD over an area. Formerly called *lines of force*. **2.** a liquid or paste used to promote easy flow of SOLDER or brazing metal over a metal surface (see SOLDERING).

flux density the concentration of electric or MAGNETIC FLUX per unit area, measuring the strength of a FIELD.

fluxmeter a meter for measuring MAGNETIC FLUX. The most common modern type is the HALL-EFFECT fluxmeter, which uses the effect of the magnetic field on a tiny semiconductor slab. The slab carries current, and applying a magnetic flux causes a potential difference to be generated across the slab. This potential difference is amplified and used to drive a meter that is calibrated in terms of flux or flux density.

flyback the part of a TIMEBASE waveform in which voltage or current returns to its starting state. See Fig. 33.

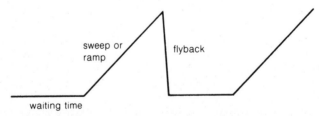

FIG. 33. **Flyback.** The flyback portion of a sawtooth wave. The diagram also shows the waiting time and the sweep or ramp portions of the waveform.

flying-spot scanner a device that enables films to be televised. The scanner consists of a high-brightness CATHODE-RAY TUBE whose moving SPOT (1) is focused onto the film. The light that passes through the film is picked up by a photocell and used to produce the output signal. After addition of blanking pulses, sync pulses, and color sync, this can be used as a standard TV signal.

flywheel 1. a rotating mechanical device that makes use of inertia to store kinetic energy. 2. *adj.* (of a SYNCHRONIZING PULSE in a TV receiver) making use of a sine-wave OSCILLATOR for generating the line timebase frequency. This oscillator uses a relatively HIGH-Q resonant circuit so that the oscillation frequency will continue unchanged even if the synchronization is irregular.

FM see FREQUENCY MODULATION.

focusing the effect of making all the electrons in a beam converge to a point. See also CONVERGENCE.

foil capacitor a CAPACITOR that makes use of metal foil as its conducting electrodes. The term usually denotes paper capacitors that consist of thin oiled paper sandwiched between metal foils. It can also denote various types of electrolytic capacitor in which the plate is of foil, such as the tantalum foil capacitor.

folded dipole a DIPOLE ANTENNA in which the rod has been folded twice. This leaves the two connections at the center as before, but reduces the IMPEDANCE of the antenna.

forced-air cooling the cooling of power components, usually transmitting tubes, by passing air from a fan over the hot material.

form factor 1. a conversion factor for calculating RECTIFIER output. The form factor for a wave is the ratio of ROOT-MEAN-SQUARE value to the rectified DC value (average value) for a half-cycle. 2. the ratio of length to diameter for a coil, used in NOMOGRAMS for finding inductance.

forward AGC a system of AUTOMATIC GAIN CONTROL used with modern transistors. These transistors are operated with a resistive load, so an increase in BIAS will reduce the collector voltage, and the characteristics of the transistors are such that this will reduce the gain. The steady voltage at the demodulator is therefore used to add to the voltage bias. Compare REVERSE AGC.

forward bias a BIAS in the direction that causes current to flow easily. The term is used mainly in referring to semiconductor junctions and other types of diodes. Compare REVERSE BIAS.

forward current the amount of current flowing in the forward, freely conducting, direction.

FORWARD DROP

forward drop the potential difference across a forward biased device such as a semiconductor diode.

forward wave 1. a wave traveling in the intended direction. For example, when a transmitter is connected to an ANTENNA system, the forward wave travels from transmitter to antenna. There may be a reflected backward wave from antenna to transmitter that will then result in STANDING WAVES being set up. 2. the input wave in a TRAVELING-WAVE TUBE.

Foster-Seeley discriminator a form of DISCRIMINATOR (1) circuit that was formerly used in FM receivers of high quality.

Fourier analysis a mathematical method of determining the frequency content of a signal. Any NONSINUSOIDAL waveform can be shown to consist of a mixture of frequencies at various amplitudes and phases. Fourier analysis enables the amplitude and phase of each component of a signal to be calculated.

four-layer diode a semiconductor device with four alternating layers of p-type and n-type material. See THYRISTOR, SILICON-CONTROLLED SWITCH.

four-terminal resistor a form of standard RESISTOR for use with POTENTIOMETER measurements. Two thick connectors are provided for passing current through the resistor. Two thin connectors are then used for connecting the resistor to the potentiometer for POTENTIAL DIFFERENCE measurements. This greatly reduces any errors that would be caused by the resistance of contacts.

frame a set of two television fields. See RASTER.

frame aerial an ANTENNA wound in the form of a large loop, used in direction finding.

Franklin oscillator a form of OSCILLATOR that employs a two-stage amplifier, with inversion in each stage, and one tuned circuit.

free electron an electron free to move in a solid, liquid, or gas. See CONDUCTION ELECTRONS.

free-field calibration the CALIBRATION of a MICROPHONE in terms of output per unit of sound pressure. The calibration has to be carried out so that the presence of the microphone does not disturb the sound-wave pattern—the free field. If this is not possible, the results have to be adjusted to allow for the field disturbance.

free oscillation an OSCILLATION not supplied with energy after an original impetus. Free oscillations will die away at a rate that depends on the amount of damping (see DAMPED) in the circuit.

free space any space, such as a vacuum, that has unity values of RELATIVE PERMITTIVITY and RELATIVE PERMEABILITY. In practical terms, air approximates conditions of free space.

frequency the rate at which a waveform action repeats, measured in HERTZ.

frequency analyzer a measuring device that can measure the amplitude and phase of the frequency components of a signal, including FUNDAMENTAL FREQUENCIES and HARMONICS. See FOURIER ANALYSIS.

frequency band a range of FREQUENCIES, usually RF, used for a particular purpose in telecommunications. There are, for example, frequency bands allocated by international agreement for medium wave broadcasting, television, mobile radio, and the like.

frequency changer the important mixer stage in a SUPER-HETERODYNE receiver. In this stage, the incoming and variable signal frequency is converted to a fixed intermediate frequency for subsequent amplification.

frequency compensation the design of a circuit to enable its use for wideband purposes. See COMPENSATED ATTENUATOR.

frequency deviation the change of CARRIER FREQUENCY caused by FREQUENCY MODULATION.

frequency discriminator a CIRCUIT that produces an output voltage proportional to input frequency over a limited range. Discriminators are used in FM demodulation and in frequency controllers. Circuits such as the RATIO DETECTOR and the FOSTER-SEELEY DISCRIMINATOR have been almost completely replaced by pulse-counting integrated circuits, such as the PHASE-LOCKED LOOP.

frequency divider a CIRCUIT whose output is at a lower frequency than that of the input, equal to the input frequency divided by an integer. Dividers generally make use of high-speed digital COUNTER integrated circuits.

frequency doubler a circuit whose output is at twice the frequency of the input. See FREQUENCY MULTIPLIER.

frequency meter an instrument used to measure FREQUENCY. Today, this consists of a counter and a time standard, which effec-

FREQUENCY MODULATION (FM)

tively counts the number of cycles of the input in one second. For microwave frequencies, wavelength rather than frequency is measured, using calibrated CAVITY RESONATORS.

frequency modulation (FM) the MODULATION of a wave by altering its FREQUENCY rather than its AMPLITUDE. The amplitude of the modulating wave corresponds to the frequency of the carrier. The maximum change of frequency permitted by the specification for the system is called the FREQUENCY DEVIATION. The main advantage of the system is that fluctuations of carrier amplitude caused by fading or interference should not affect the demodulated output at the receiver. See also PHASE MODULATION.

frequency multiplier a device whose output is at a FREQUENCY that is an integer multiple of the input frequency. Frequency multiplication is used in transmitters, particularly as a way of enabling a single crystal oscillator to provide several different control frequencies in a number of different bands.

The basis of frequency multiplication is an underbiased stage, whose output is rich in harmonics. The required harmonic is selected by making the load a resonant circuit of the desired frequency. The most common multiplication ratios are two (FREQUENCY DOUBLER) and three (frequency tripler), because the amplitude of harmonics at higher ratios is generally insufficient for reliable use.

frequency pulling an alteration in the FREQUENCY of an OSCILLATOR because of the electrical effect of a circuit connected to it. Applies to alteration of frequency caused by loading, because the frequency of oscillation is affected by the dynamic resistance of the oscillator tuned circuit, and the load is connected in parallel with this dynamic resistance, thus reducing the overall value. Frequency pulling is a particular problem in BEAT-FREQUENCY OSCILLATORS, which are intended to produce low frequency outputs.

frequency range the set of frequencies over which a device or circuit is intended to operate.

frequency standard 1. a stable oscillator whose output is used to calibrate other instruments. Today, low cost rubidium MASERS can be bought for this purpose. **2.** the standard transmissions from the transmitter of the Environmental Research Laboratories, in Boulder, Colorado, which are used to check oscillator frequencies in the US.

fringe an area at a great distance from a transmitter in which reception is just barely usable.

front porch a part of a TV SYNCHRONIZING PULSE consisting of a short interval of BLACK LEVEL that immediately precedes the line synchronizing pulse.

frying 1. the sizzling sound that usually indicates ARCING in a high voltage circuit. **2.** noise in an audio circuit caused by poor contacts.

fsd see FULL-SCALE DEFLECTION.

full adder a LOGIC circuit that carries out the complete addition action. For two-bit operation, a full adder needs three inputs, consisting of the two bits to be added, plus any carry bit from a previous operation. The output consists of the sum digits and the carry bit for use by subsequent devices. See ADDER, HALF-ADDER.

full-scale deflection (fsd) the deflection of the needle of a meter movement to the last figure on the dial.

full-wave dipole a DIPOLE antenna whose total length is equal to a full wavelength of the signal frequency.

full-wave rectifier a RECTIFIER circuit that produces a positive half-wave peak in one direction for each half of an AC wave. This is carried out by means of a bridge rectifier circuit. See Fig. 34.

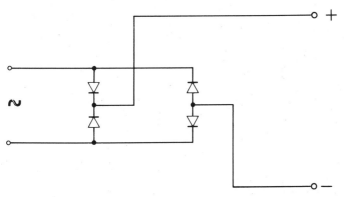

FIG. 34. **Full-wave rectifier.** The circuit uses a diode bridge.

function generator a type of signal GENERATOR that can produce a set of different waveforms.

fundamental frequency or **basic frequency** the lowest frequency of repetition in a wave that can also include HARMONICS. See also FOURIER ANALYSIS.

fuse a safety device consisting of a short length of metal wire made from a material with a low melting point. Excessive CURRENT flowing through the fuse will melt the metal, thus interrupting the circuit. The fuse then has to be replaced after the fault that caused the excessive current has been located and repaired. See also CIRCUIT BREAKER.

G

gain or **amplification** the ratio of signal output to signal input for an amplifier. The ratio is often given in decibels, using 10 log (out/in) for power ratios and 20 log (out/in) for voltage ratios. The gain of an ANTENNA is the ratio of the signal power transmitted or received in the intended direction compared with the signal to or from a simple DIPOLE. Compare ATTENUATION.

gain control a POTENTIOMETER (1) or switch for varying signal amplitude.

gallium arsenide a compound SEMICONDUCTOR. Unlike silicon or germanium, gallium arsenide is not an element. It is almost impossible to dope by DIFFUSION and modern methods of DOPING rely on ION IMPLANTATION. The doped material has low resistivity, but its reversed biased junctions are of high resistance. At present, gallium arsenide is used mainly for microwave diodes, and for high-speed field-effect transistors that can be used at UHF.

ganged circuits a set of CIRCUITS in which VARIABLE CAPACITORS, INDUCTORS, or POTENTIOMETERS are mechanically linked and operated with a single control. The ganging of variable capacitors in a SUPERHETERODYNE receiver, for example, enables both tuning of the oscillator and altering of the antenna frequency circuits in one step.

ganging the mechanical connection of adjustable COMPONENTS so that several can be operated by a single control.

gap a narrow break in a CIRCUIT. A *magnetic gap* consists of a narrow piece of nonmagnetic material placed in a magnetic circuit. This is put in to prevent SATURATION of the magnetic material. A SPARK GAP is often included in high-voltage circuits. This consists of two metal conductors close together, one at high voltage and the other grounded. The principle is that a voltage surge will cause energy to be lost in the form of a spark rather than in damaging components.

gas breakdown see AVALANCHE.

gas discharge the flow of current in an ionized gas (see IONIZATION). Gases do not conduct current unless they are ionized, and a high electric field must be used to start ionization. The voltage required to start ionization will be higher in the absence of light unless the gas in the tube is slightly radioactive.

One former use of gas-discharge tubes was as voltage reference standards, since the voltage drop across a conducting gas is fairly constant. During the action of a gas-filled tube, the gas can be seen to glow, and this leads to applications of the effect as display devices. See DISCHARGE.

gas-filled tube any type of tube that contains low-pressure gas rather than a vacuum. GEIGER-MÜLLER TUBES are gas-filled, as were the thyratrons and ignitrons once used as electronic relays. Gas-filled tubes were also used as voltage-reference devices and for display.

gate 1. the controlling electrode of a FIELD-EFFECT TRANSISTOR. 2. a CIRCUIT that uses one WAVEFORM to control the passage of another. In analog circuits, a gate will pass an input signal for a period of time that is determined by a gating signal, usually a square wave. Digital circuits use gates to carry out logic actions. See AND GATE, OR GATE, EXCLUSIVE-OR GATE, NOT GATE.

gate array a type of INTEGRATED CIRCUIT that consists of many separate gates. If these gates are not connected to each other, the array is called an UNCOMMITTED LOGIC ARRAY. See EPLD.

gate expander a CIRCUIT connected to the INPUT of a logic gate to enable further inputs to be connected. See FAN-IN, FAN-OUT.

Geiger-Müller tube a type of GAS-FILLED TUBE used to detect ionizing radiation. The tube contains a low-pressure gas, and a trace of deionizing material such as bromine. Two electrodes in the tube are kept at a potential difference that is lower than needed to start a discharge. When an ionizing particle enters the tube, gas is ionized, but the effect of the deionizing material is to cause recombination of these ions when the particle passes out of the tube. The tube therefore passes current only for a brief interval, one pulse for each particle. See also DEAD TIME.

generator 1. a DEVICE that can supply electrical power, such as a dynamo and an alternator, both of which convert mechanical energy into electrical energy. 2. any signal source that can be represented as a voltage or current generator in an equivalent circuit.

germanium the first element to be used for manufacturing transistors. It has been almost completely superseded by SILICON in this technology.

getter a material for removing the last traces of gas from a vacuum tube. A getter consists of a thin film of a metal such as barium. This reactive metal will absorb any traces of gas produced as the tube is used. The use of getters is confined to the smaller receiving vacuum tubes and to CATHODE-RAY TUBES because transmitting vacuum tubes run at a temperature that would vaporize the getter. See also OUTGASSING.

ghost see ECHO.

giga- *combining form* denoting a billion.

glass electrode a device for measuring the acidity (pH) of solutions, consisting of a wire connection to a conducting liquid inside a thin glass bulb. The voltage from a glass electrode needs to be amplified before it can be measured.

glow lamp a GAS-FILLED TUBE used as an indicator or voltage regulator.

Gm see MUTUAL CONDUCTANCE.

gold bonding a method of making a JUNCTION between N-TYPE germanium and a gold wire. The method was formerly the only one available for manufacture of high-speed diodes.

graph recorder a form of CHART RECORDER.

graphic symbol a symbol used to represent a component in CIRCUIT DIAGRAMS.

grass see GROUND CLUTTER.

grid an ELECTRODE that controls ELECTRON flow in a vacuum tube. The type used in CATHODE-RAY TUBES consists of a small aperture in an electrode shaped like a top hat. In a transmitting tube, the grid is a winding of molybdenum wire surrounding the hot FILAMENT.

grid bias the voltage applied between CATHODE and GRID of a vacuum tube to control electron flow. The BIAS is normally applied with the grid negative, and at a sufficiently large negative potential the electron stream will be cut off.

grid dip a form of passive TELEMETRY in which inductive COUPLING is used to transmit information over distances of up to 50 cm. The name arises from a circuit now obsolete that was used to detect the frequency of a passive resonant circuit by its effect on the grid current of a vacuum tube oscillator tuned to the same frequency. Grid-dip telemetry is used for biomedical work in which measurements have to be made by sensors implanted within a patient, where no physical connections are possible.

grid emission the emission of electrons from an overheated GRID usually in a transmitting tube used near its maximum ratings. The electrons from the grid will be accelerated to the anode and will not be controlled by the grid bias. This can result in more overheating, causing failure of the tube. Grids are commonly coated with graphite to reduce the emission of electrons and to help in dissipation of heat by radiation.

grid modulation a method of achieving AMPLITUDE MODULATION in vacuum tube transmitting equipment. The final power amplifier tube has its BIAS applied through a transformer winding to which the modulating frequency is applied. The advantage of this form of modulation is that the power required is relatively low. The disadvantage is that the power output stage has to be reasonably linear in operation, which is not the case with the common CLASS C design.

ground any zero-voltage point. The earth is taken as being of zero voltage, because its potential is not greatly affected by small changes of charge.

ground absorption the loss of radiated signal power to ground. If the ground wave is absorbed rather than reflected, the range of the transmitter will be fairly short.

ground bonding the making of a connection to ground. This is often done by burying a metal plate in the earth and connecting to the plate. An alternative is to use metal waterpipes, but care must be taken to connect to a pipe that passes through the ground.

Heating pipes and hot water pipes should not be used. The use of plastics for water pipes and other pipes means that great care is needed in grounding equipment, particularly transmitters.

Electrical circuits are commonly multiply grounded, meaning that all pipework and other metal are bonded to a common ground that is also used for the AC supply ground.

ground capacitance the CAPACITANCE of any point in a circuit to ground. If this capacitance is variable, it may lead to tuning errors in RF receivers.

ground clutter RADAR reflections from irregular ground. These cause a fuzzy screen display that looks like grass.

ground current a CURRENT flowing to or through the earth. A current will flow to the earth if an insulation fault develops in grounded equipment. This type of fault should cause fuses to blow, or contact breakers to open. Currents can also flow through the earth, and may cause electrolytic corrosion of metals, including grounding plates.

grounded-base connection see COMMON-BASE CONNECTION.

grounded-collector connection see EMITTER FOLLOWER.

grounded-emitter connection see COMMON-EMITTER CONNECTION.

ground fault an insulation breakdown that causes current to flow to ground.

ground loop a cause of unwanted FEEDBACK of signal that is caused by using separated ground connections in a circuit. The currents flowing between the two ground points result in a potential difference that provides the unwanted feedback signal.

ground plane a sheet of metal that is grounded or, particularly in a battery-operated circuit, used as a ground. The surrounding metal on a printed circuit board is often used as a form of ground plane.

ground potential the zero potential (see ELECTRIC POTENTIAL). All potentials on ground are measured relative to the potential of the earth, which does not change rapidly.

ground return 1. ground echoes that cause GROUND CLUTTER. 2. the return half of a CIRCUIT in which a wire is used for the first half. A ground return was used in early Morse code TELEGRAPHY.

ground wave a wave that has been reflected from the earth. If this happens to a radar echo, the distance measurements for the signal will be incorrect. Compare SKY WAVE.

group velocity the velocity of a group of electromagnetic waves in a medium other than a vacuum. This velocity is always less than the velocity of the waves in a vacuum. Individual waves within the group *appear* to travel with a PHASE VELOCITY greater than the velocity of the waves in a vacuum.

grown junction an old method of producing a semiconductor JUNCTION by adding DOPING impurities to the molten material from which a crystal is being pulled.

guard band 1. an unused FREQUENCY BAND. The purpose of this is to avoid the chance of crosstalk between two adjacent bands. 2. a grounded metal strip in standard capacitors that prevents the spread of an electric field.

Gunn diode a microwave DIODE used as an oscillator in microwave applications. It has replaced KLYSTRONS for many applications. The Gunn diode is made of GALLIUM ARSENIDE and emits microwave radiation when a large electric field is placed across the material.

Gunn effect the generation of MICROWAVE OSCILLATIONS from a GALLIUM ARSENIDE diode.

gyrator a MICROWAVE circuit COMPONENT. Its action is to pass signals in one direction, but reverse the phase of signals traveling in the other direction.

H

halation a fuzzy glow around the SPOT on a CATHODE-RAY TUBE. The effect is noticeable on oscilloscope and radar tubes and is caused by internal reflections of light within the thick glass at the front of the screen.

half-adder a simple adding circuit for two binary bits. Only two inputs are needed, because no carry input or output is provided. A half-adder can be constructed from any standard GATE type, such as NAND or NOR. See Fig. 35. See also ADDER, FULL ADDER.

half-duplex a line communication system that allows signals to pass in either direction, but not at the same time. Compare DU-PLEX, SIMPLEX.

half-power point or **-3dB point** a graph point used in determining the Q FACTOR of a RESONANT CIRCUIT. The frequency limits at which the output power is equal to half of the maximum power output.

half-wave dipole a DIPOLE ANTENNA whose length is half of the radiated wavelength.

half-wave rectifier a RECTIFIER, for example, a single diode that passes one half of an AC voltage and blocks the other half, which has the opposite polarity.

Hall effect a VOLTAGE generated by the effect of a magnetic field on a slab of material that is carrying current. SEMICONDUCTORS give particularly large Hall effect voltages, and the effect is used to measure magnetic fields (see FLUXMETER).

Hall probe the probe of a HALL EFFECT fluxmeter, containing the semiconductor slab in which the Hall effect occurs.

ham a slang term for amateur radio enthusiast.

Hamming code a complicated error correcting code used in data transmission. Hamming code is used to correct TELETEXT data signals.

HANOVER BARS

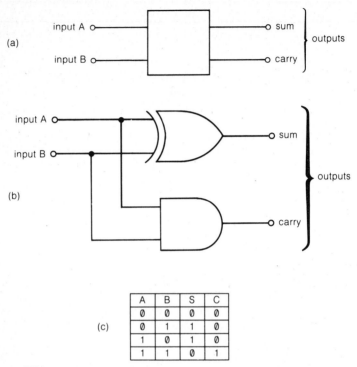

FIG. 35. **Half-adder.** (a) The two-bit half-adder symbol. (b) Practical implementation. (c) Truth table.

Hanover bars a fault condition of the PAL type color TV system in which the picture appears broken into colored horizontal bars.

hard disk a magnetic data-storage device for computers, in which a rigid disk coated with a ferromagnetic material is spun at high speed. The read-write head floats above the disk on an air cushion. Extremely fast access times can be achieved by this system.

While its cost originally restricted its use to MAINFRAME computer systems, low-cost versions such as the WINCHESTER DISK are now available.

hard tube an electronic tube that uses a vacuum, as distinct from a GAS-FILLED TUBE.

hardware the electronic components of a computer system, as distinct from program instructions. Compare SOFTWARE.

hardwire assemble a CIRCUIT by wiring from component to component rather than by using a printed circuit.

harmonic a part of a signal whose frequency is an integer multiple of the FUNDAMENTAL FREQUENCY of the signal. Any periodic waveform other than a sine wave will contain harmonics. Compare SUBHARMONIC.

harmonic generator a CIRCUIT that deliberately generates HARMONICS. It can use reverse-biased transistors, diodes, or other NONLINEAR elements. The aim is to steepen the sides of any sine wave and so result in the generation of harmonics.

Hartley oscillator or **tapped-coil oscillator** an oscillator circuit that uses a tapped inductor to obtain a feedback signal. See Fig. 36. See also COLPITTS OSCILLATOR.

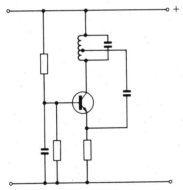

FIG. 36. **Hartley oscillator.** The figure shows one form of the Hartley oscillator circuit, using a tapped inductor for feedback.

H-attenuator an ATTENUATOR of symmetrical construction intended for attenuating two signals by an equal amount. These signals may be the inputs or outputs of a BALANCED AMPLIFIER. See Fig. 37.

head a reading or writing recording device. See READ/WRITE HEAD, RECORDING OF SOUND, SOUND REPRODUCTION.

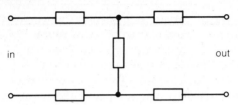

FIG. 37. **H-attenuator.** An example of an H-attenuator, using resistors. The design is symmetrical and is intended for balanced signals.

heat detector any device that senses temperature change. Examples include THERMOCOUPLES, THERMISTORS and PYROELECTRIC generating materials.

heater 1. a vacuum tube filament or other heating element. 2. the molybdenum spiral that is placed inside the CATHODE of an indirectly heated vacuum tube, and insulated from it. 3. another name for (a) a WIRE-WOUND RESISTOR in a crystal oven, (b) a THERMOSTAT accelerating heater.

heating effect the effect of CURRENT I driven by a voltage V passing through a RESISTANCE R. The heat dissipation rate P in watts may be found from $P = IV$ or $P = I^2R$.

heating time the time needed for a hot cathode device to reach operating temperature.

heat sink a massive metal slab to which power transistors can be bolted. The purpose of a heat sink is to spread the conducted heat from the collector of a transistor over a large mass and a large area, and ultimately to the air.

    The transistor will still have to be operated in such a way that the rate of power dissipation does not exceed the rate of heat dissipation.

Heaviside layer see E LAYER.

height control the TELEVISION RECEIVER control for amplitude of frame scan. This has the effect of altering picture height. Modern receivers either dispense with this control or use a preset POTENTIOMETER.

helical antenna an ANTENNA that consists of a wire in a corkscrew shape. The radiated wave is elliptically POLARIZED (1).

helical potentiometer (helipot) a POTENTIOMETER in which the element is in helical form. This means that several turns of the actuating shaft are needed to wind the contact from one end of the element to the other. Such a potentiometer, using anything from 3 to 50 turns from one stop to the other, has very high resolution (see RESOLUTION OF POTENTIOMETER), and is used in measuring instruments.

helipot see HELICAL POTENTIOMETER.

Helmholtz coils a pair of identical coils placed on the same axis and separated by a distance equal to coil radius. The field in the space between the coils is fairly uniform.

henry the unit of INDUCTANCE self or mutual. Symbol: H. The henry is the unit equivalent to volt-seconds per ampere. Named for Joseph Henry (1797—1878), US physicist.

hertz (Hz) the unit of FREQUENCY equal to one cycle per second. Symbol: Hz. Named after Heinrich Hertz (1857—94), German physicist.

heterodyne 1. (of a circuit) producing a BEAT FREQUENCY by mixing an incoming frequency with a locally generated oscillation that is at a slightly different frequency. **2.** *vb.* mix two frequencies together to produce a beat frequency, as in a SUPERHETERODYNE RECEIVER or a HETERODYNE WAVEMETER.

heterodyne wavemeter a form of FREQUENCY METER that operates with a calibrated oscillator. The oscillator frequency is varied until beats (see BEAT FREQUENCIES) of a particular frequency, often zero, are detected. The input frequency is then read from the calibrated scale.

heterojunction a JUNCTION (2) between different semiconductor materials, or between a semiconductor and a metal.

hexadecimal scale the number scale that uses 16 digits. The letters A to F are used to represent the decimal numbers 10 to 15 in this scale. For example, the hexadecimal number A5 is the decimal number 165.

HF see HIGH FREQUENCY.

hi-fi *abbreviation for* high-fidelity sound. Hi-fi covers all aspects of RECORDING OF SOUND and SOUND REPRODUCTION with the aim of securing the most realistic reproduction possible. The term is often used indiscriminately to denote stereo audio systems that are of noticeably low fidelity.

high frequency (HF) 1. the FREQUENCY BAND that covers the frequencies between 3 and 30 megahertz approximately. 2. any frequency higher than midrange audio frequency.

high-logic level the actual voltage level taken to represent the logic 1. For most modern transistor-transistor LOGIC CIRCUITS this means any voltage between +3.5 and +5.0 volts.

high-pass filter a FILTER whose PASSBAND starts at a high frequency and continues indefinitely.

high-Q (of an inductor or tuned circuit) having a large ratio (100 or more) of inductive IMPEDANCE to resistance at the RESONANT FREQUENCY.

high tension (HT) an obsolete term for high VOLTAGE usually taken to mean voltages in the range of 100 to 500V.

high-voltage test a method of testing INSULATION. The test consists of applying higher than normal voltages across the insulation, often in the form of pulses. The quality of the insulation is measured by the current that passes during each pulse.

hiss the particularly obtrusive form of NOISE from recorded tape caused by random magnetization of particles on the tape. Tape hiss is particularly noticeable on narrow tapes operated at low speeds. This makes some form of COMPANDER system, such as DOLBY or DBX essential for cassette systems if reasonably good reproduction is needed.

holding current the minimum CURRENT needed to maintain conduction of a THYRISTOR. The current that flows between the anode and cathode of a thyristor serves also to keep the thyristor switched on. If this current falls below a minimum holding level, the thyristor will switch off. Too low a voltage in the forward direction will also allow the thyristor to switch off.

hole the space left in the atomic arrangement of a crystal when an ELECTRON is removed. This space is mobile and behaves as if it carries positive charge.

hole conduction the CONDUCTION in a metal or a semiconductor caused by HOLE movement in the direction of a positive field.

hole current the amount of CURRENT that is due to HOLE movement. In several metals, a substantial amount of the current that can flow is conducted by hole movement. In p-type semiconductors, almost all of the current is conducted by hole movement.

hole trap a fault in a semiconductor CRYSTAL that can stop the movement of HOLES by allowing recombination with an electron.

homing beacon see BEACON.

homopolar generator a GENERATOR that uses a rotating metal disk rather than a coil. The homopolar generator delivers low-voltage DC without use of a COMMUTATOR.

honeycomb coil an old-fashioned type of coil construction that reduces STRAY CAPACITANCE between the turns of the coil.

hookup or **patch** 1. a radio or telephone circuit. 2. a temporary circuit for assessment or measurement. 3. a connection between a radio link and a telephone system.

horizontal blanking a SIGNAL applied to the grid of a receiver or camera CATHODE-RAY TUBE to cut off the beam and thus prevent signals from being generated or displayed during the LINE FLY-BACK period.

horizontal hold the control over SYNCHRONIZATION of the HORI-ZONTAL TIMEBASE of a TELEVISION RECEIVER.

horizontal polarization a property exhibited by an electromagnetic wave in which the electric field vector is horizontal. A wave transmitted from a horizontal antenna will be horizontally polarized. The direction of polarization of a wave is by convention taken to be the direction of the electric-field component of the wave.

horizontal sweep the movement of an ELECTRON BEAM horizontally across the face of a CATHODE-RAY TUBE. This is the result of applying a ramp waveform of voltage to horizontal deflection plates in an INSTRUMENT TUBE or a ramp waveform of current to the horizontal deflection coils of a PICTURE TUBE.

horizontal timebase the circuit that generates the horizontal TIMEBASE waveform.

horn antenna a MICROWAVE antenna consisting of a horn that starts with the waveguide and flares out in each direction.

horn loudspeaker or **horn-loaded loudspeaker** the most efficient form of LOUDSPEAKER used as a comparison standard. The horn loudspeaker consists of a DRIVER unit, which can be a conventional electromagnetic unit with a small cone. The speaker walls then follow the pattern of an exponential curve, increasing in area.

This loudspeaker is a relatively efficient converter of electrical energy into sound energy, 4% typically, but requires a lot of space. Care must be taken that the material used for the walls of the horn does not resonate within the audible frequency range.

horseshoe magnet a permanent MAGNET whose shape is like that of a horseshoe. This means the poles of the magnet are close, causing an intense field across the air gap.

hot-carrier diode see SCHOTTKY DIODE.

hot cathode a CATHODE that emits electrons into a vacuum when heated. See also COLD-CATHODE EMITTER.

hot dip or **tinning bath** a method of coating metals with other metals by dipping them into a bath of molten metal. The method is used to coat PRINTED CIRCUIT BOARDS with solder.

hot electron a high energy electron.

hot spot a part of an ELECTRODE that is overheated. Hot spots become particularly troublesome in vacuum devices if they begin to emit electrons as a result of their high temperature. A hot spot in a semiconductor can cause diffusion of impurities, thus altering the characteristic of the material.

hot-wire meter a current-measuring instrument that operates on the hot-wire principle. In a typical hot-wire meter, a wire is held between supports, and a tensioning spring is attached to the center of the wire. The attachment to the needle movement is also made at the center. When the wire is heated by passing current it will expand. The resulting sag at the center of the wire is mechanically amplified by levers and used to drive the needle. Hot-wire instruments, though not very accurate, are useful for indicating true ROOT-MEAN-SQUARE values of high-frequency currents.

howl an unwanted audio OSCILLATION. Howl is caused by unwanted FEEDBACK and this can be in audio-amplifier stages, in radiofrequency stages, or as a result of acoustic feedback, for example, from a loudspeaker to a microphone or phonograph pickup.

h-parameter *abbreviation for* hybrid PARAMETER. Hybrid parameters are sets of CHARACTERISTIC figures that have different units. When graphs are drawn of component or circuit characteristics, the slope of each graph will yield a parameter figure that will have units such as resistance, conductance, or gain. Hybrid parameters form sets of units that are not consistent, that is, they are not all resistances, nor all ratios, but a mixture of different types.

H-plane the direction of the magnetic field portion of a MICRO-WAVE signal. See WAVEGUIDE.

HT see HIGH TENSION.

hue the spectral color of a portion of a picture. In COLOR TELEVISION applications, a given color has two measurable quantities, hue and SATURATION. Hue refers to the color itself, saturation to the ratio of pure color to white.

hum a low frequency signal, usually picked up from AC power supplies. The picking up of hum is a problem for designers of high-gain audio equipment. It cannot be eliminated by using simple FILTERS because the action of a RECTIFIER generates HARMONICS of the hum frequency, and these harmonics radiate.

hum modulation the MODULATION of signal AMPLITUDE by low-frequency hum.

hunting an OSCILLATION in a control SYSTEM that is, in any system in which electronic circuits control mechanical or other nonelectronic actions and in which a FEEDBACK CONTROL LOOP is used. Hunting is the consequence of poor design in which a NEGATIVE FEEDBACK loop with insufficient damping (see DAMPED) has been applied to an unsuitable system. The result is that the feedback overcorrects, and the system never becomes stable.

hybrid IC a type of INTEGRATED CIRCUIT that mixes COMPONENTS on the chip with components separately connected. The added components, which are welded to bonding pads, are components that cannot be manufactured easily in IC form, for example, capacitors, inductors, and quartz crystals. Hybrid ICs are used mainly in military applications.

hybrid junction or **Magic-T junction** or **rat-race junction** a method of connecting WAVEGUIDES so as to change signal direction and avoid reflected signals.

hybrid-pi equivalent an old form of EQUIVALENT CIRCUIT for a TRANSISTOR. This has seldom been used since silicon transistors became predominant, because simpler equivalents can now be used.

hysteresis an effect in which the direction of change of a quantity

must be considered as well as size of change. See HYSTERESIS
LOOP.

hysteresis distortion a form of signal distortion that is caused by
HYSTERESIS.

hysteresis loop the shape of a graph of any quantity that exhibits
hysteresis. The best-known example is MAGNETIC HYSTERESIS.
Such a graph has a loop shape, because the path taken when one
plotted quantity is increased is not the same as the path taken
when the quantity is decreased. Only the ends of the loop are
common to the two paths. The area of the loop usually indicates
the amount of energy required during each cycle.

hysteresis loss the loss of energy resulting from tracing out a
graph that contains a HYSTERESIS LOOP. The amount of energy
lost is represented by the area of the hysteresis loop.

Hz see HERTZ.

I

IC see INTEGRATED CIRCUIT.

ideal bunching the gathering of moving electrons into tightly packed groups, with no electrons between the groups. See BUNCHER, KLYSTRON.

ideal crystal see PERFECT CRYSTAL.

ideal transformer see PERFECT TRANSFORMER.

idle component see REACTIVE CURRENT.

IEE *abbreviation for* Institution of Electrical Engineers, the senior electrical-engineering institution in Britain.

IEEE *abbreviation for* Institute of Electrical and Electronic Engineers, the US institute for professional engineers in all branches of electrical technology.

IF see INTERMEDIATE FREQUENCY.

IFF *abbreviation for* identification friend or foe, a form of RADAR in which a coded radar signal is decoded and correctly returned by a friendly aircraft. See INTERROGATING SIGNAL, TRANSPONDER, SECONDARY RADAR.

IF strip the part of a SUPERHETERODYNE RECEIVER that operates at the INTERMEDIATE FREQUENCY. This consists of all the stages from the MIXER to the demodulator, along with the AGC connections.

IGFET see FIELD-EFFECT TRANSISTOR.

ignitron an old form of mercury-vapor controlled RECTIFIER for very high voltage and large currents.

image converter or **image intensifier** a device that achieves light amplification or frequency changing by use of a PHOTOCATHODE and a PHOSPHOR operated with a large potential difference between them. An image that may be of ordinary light, ultraviolet light, or infrared light, causes electrons to be emitted. These electrons are accelerated by the electrodes, and their relative positions are maintained until they strike the phosphor. The electrons that strike the phosphor form a bright image. This image will be a visible light image, a copy of the original image, which may be too dim to see, or it may be composed of invisible wavelengths. Image converters and intensifiers are used in x-ray systems, making it possible for x-ray dosages to be greatly reduced, and in infrared night-sighting systems.

image frequency the frequency of a signal into a SUPER-HETERODYNE RECEIVER that will provide the same INTERMEDIATE FREQUENCY as the correct frequency to which the receiver is tuned. For example, if the receiver is tuned to 98MHz, using a 10.7MHz IF, then the oscillator frequency will be 108.7MHz. An input frequency of 119.4MHz, however, will also produce an IF of 10.7MHz (by subtracting the oscillator frequency), and this is the image frequency. If the IF is suitably chosen, the image frequency should be spaced sufficiently far from the desired frequency to be rejected by the RF tuning circuits. When a low frequency IF is needed, the DOUBLE-SUPERHET RECEIVER principle will prevent image frequency problems.

image impedance the value of IMPEDANCE measured at the terminals of a FILTER network, that is, the values of *input impedance* with output open circuit and *output impedance* with input open circuit.

image orthicon a form of TV CAMERA TUBE that was used until color TV broadcasting began. At that point, the image orthicon was largely replaced by the VIDICON.

IMPATT diode a form of junction DIODE that oscillates when reverse-biased into AVALANCHE conditions. The oscillations are at MICROWAVE frequencies and can be tuned by placing the diode in a RESONANT CAVITY.

impedance the complex sum of AC RESISTANCE and REACTANCE. Symbol: Z; unit: ohm. If an AC signal is applied to a circuit containing reactive components, the phase of the output current need not be the same as the phase of the output voltage. The relationship between **V** (complex voltage) and **I** (complex current) cannot therefore be described by a scalar quantity such as resistance, as is the case for DC circuits. A two-part quantity such as a vector or COMPLEX NUMBER is needed, and this quantity is termed *complex impedance*. See Fig. 38.

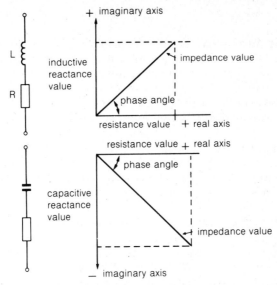

FIG. 38. **Impedance.** See this entry.

impedance matching a means of ensuring that the IMPEDANCES of two circuits to be connected are equal. Impedance matching is particularly important for RF power circuits and TRANSMISSION LINES. The most efficient transfer of power from an AMPLIFIER to a LOAD occurs when the output impedance of the amplifier equals the input impedance of the load. In a transmission line, impedance matching prevents reflections that would otherwise occur at the point where unequal impedances were connected, and the reflections would lead to losses. FILTER sections that are to be connected together will operate correctly only if the impedances are matched.

impregnated cable a cable using paper or cloth insulation impregnated with oil.

impulse 1. a sudden rise and fall of voltage or current. Impulses are generated by switching, particularly in inductive circuits, and also by spark discharges. **2.** in signal analysis, a theoretical concept denoting a pulse of infinite height and zero width.

impulse generator a GENERATOR that provides an impulse voltage or current, which may be used, for example, for testing insulation.

impulse noise a type of radio frequency NOISE that arises from SPARKS. The spark ignition system of gasoline engines is a common source of impulse noise. Another source is sparking contacts in thermostats, for example, in refrigerators and heating systems. Impulse noise is particularly troublesome to VHF transmissions, particularly FM and TV transmissions, which have a broad bandwidth.

impurities the atoms contained in a semiconductor CRYSTAL that are not of the semiconductor element or compound. Impurities may exist because they cannot be removed, or because they have been deliberately added, in which case the impurities are referred to as DOPING impurities.

incandescent lamp a lamp whose light is generated by a hot object, usually a metal filament.

incremental resistance or **slope resistance** or **differential resistance** the slope of the graph of voltage plotted against current for a component or circuit. This corresponds to the EQUIVALENT RESISTANCE for a small signal applied with the bias set to the same voltage and current levels.

indicating instrument an instrument that indicates presence of a current or voltage, but does not measure its value.

indicator tube a type of miniature CATHODE-RAY TUBE. Typically, indicator tubes are built into equipment so that WAVEFORMS can be checked without having to connect up a separate OSCILLOSCOPE.

indirect wave a radio wave that reaches a RECEIVER by reflection from, for example, the IONOSPHERE.

induce produce an electrical effect at a distance. An electrostatic CHARGE can be induced in any object by bringing up another charge. A MAGNETIC FLUX can be induced in a magnetic material by bringing up a magnet or a coil that is carrying current.

induced current the CURRENT that flows in a CONDUCTOR because the magnetic field around the conductor is cutting through lines of magnetic flux.

induced noise the NOISE that is induced from another circuit. For example, HUM in an audio circuit may be induced from a power line transformer by way of an GROUND LOOP.

inductance the quantity that relates MAGNETIC FLUX to the CURRENT that causes it. Symbol: L; unit: henry. Inductance is of two types, SELF-INDUCTANCE as applied to a single inductor such as a coil, and MUTUAL INDUCTANCE.

induction the process of inducing MAGNETISM or CHARGE.

induction compass a miniature AC GENERATOR that acts as a magnetic compass. A coil rotates, and the peak output is obtained when the axis of the coil, at right angles to the axis of rotation, faces along the magnetic north-south axis.

induction flowmeter a method of measuring the flow rate for a conducting liquid. The flow of the liquid is at right angles to a MAGNETIC FIELD and a small but measurable voltage is generated or induced across the liquid.

The importance of the method is that there need be no interference with the flow of the liquid, and the method is equally applicable to very hot or very cold liquids, or to ionized gases (see IONIZATION). The principle is also used in the magnetohydrodynamic generator as a method of generating power.

induction furnace or **induction heater** a device for melting metals by making the metal a part of the secondary winding of a TRANSFORMER. This is used for large-scale heating. For small samples, an EDDY CURRENT HEATER is preferable.

induction heater see INDUCTION FURNACE, EDDY CURRENT HEATER.

induction instrument a type of MOVING-IRON METER. The STATOR coil is supplied with the alternating current to be measured. The EDDY CURRENTS that are induced in a movable conductor cause force effects that move the conductor and an attached needle against the restoring force of a spring.

induction microphone see RIBBON MICROPHONE.

induction motor a type of MOTOR that depends on INDUCTION effects. The FIELD COILS are fed with AC and induce currents in the ROTOR coils, which may be metal bars. The forces between the two fields cause the rotor to rotate. Many types of induction motors are not self-starting and have to be spun to operating speed by an auxiliary motor.

inductive load a LOAD in which the phase of voltage leads the phase of current.

inductor a COIL loop, or wire used for its inductance value.

infrared the range of radiated waves with WAVELENGTHS of about 800 nm to 1 mm. These are the wavelengths that are longer than red wavelengths in the visible spectrum. These waves are emitted mainly by hot objects, so their main effect on any objects they strike is to heat them.

infrasonic see SUBSONIC.

inhibit suppress an OUTPUT or prevent an INPUT from having any effect.

inhibit pulse a PULSE applied to a magnetic CORE to prevent writing. See also COINCIDENT HALF-CURRENT SWITCHING.

injection 1. the application of a SIGNAL particularly a test signal, to a CIRCUIT. 2. the supply of a signal from an OSCILLATOR into a MIXER. 3. the introduction of CARRIERS into a SEMICONDUCTOR.

in-phase component 1. the portion of a voltage SIGNAL that is in phase with current, formerly called *active current*. 2. the portion of a current signal that is in phase with voltage, formerly called *active voltage*, that is, having a PHASE ANGLE of zero between the voltage wave and the current wave. Where a phase difference exists between voltage and current, either voltage or current can be considered to consist of an in-phase component and an out-of-phase component.

input 1. a SIGNAL taken into a component or circuit. Compare OUTPUT. 2. the connections to which the signal is applied.

input impedance the complex quantity (see COMPLEX NUMBER, IMPEDANCE) relating signal voltage amplitude and phase to signal current amplitude and phase at an INPUT.

input/output the parts of a computer circuit that deal with the INPUT (1) or OUTPUT of signals.

insertion gain the GAIN of an amplifier when it is placed between a source of SIGNALS and a LOAD. The figure is usually expressed in decibels.

insertion loss the loss of SIGNAL power due to a NETWORK placed between a source of signal and a LOAD. The figure is usually expressed in decibels.

instantaneous current the value of CURRENT in an AC circuit at some instant of a signal cycle.

instantaneous frequency the FREQUENCY of a SIGNAL at any instant. See FREQUENCY MODULATION.

instantaneous power the power transferred or dissipated in a CIRCUIT or a COMPONENT at an instant.

instrumentation the equipping of a system with instruments that measure and monitor the condition of the system.

instrument rating the limit of current or voltage that can be applied to a measuring instrument without causing damage. If no rating is known, a figure of 1.5 times FULL-SCALE DEFLECTION is often assumed.

instrument sensitivity the ratio of the number of dial scale divisions to the unit that is being measured. Examples are divisions per volt, per milliampere, and per watt.

instrument transformer a form of CURRENT TRANSFORMER used for measuring large ALTERNATING CURRENTS.

instrument tube a form of CATHODE-RAY TUBE intended as a measuring device rather than for picture display. Instrument tubes are operated with beam deflection times that may vary from seconds to fractions of a microsecond. Their size need not be large, but their sensitivity in terms of millimeters of deflection per volt of deflection waveform must be high.

Electrostatic deflection is used, and the leads to the deflection plates are often taken through the side of the tube wall to avoid excessive stray capacitance. The deflection usually takes place when the beam is moving comparatively slowly, and after deflection the beam is accelerated to its final velocity (see POST-DEFLEC-TION ACCELERATION). The face of the tube will be flat so that measurements of distance between points on the face can be made easily.

insulate prevent flow of CURRENT to or from a CONDUCTOR.

insulated gate FET(IGFET) see FIELD-EFFECT TRANSISTOR.

insulated system an ungrounded electrical supply system. Such a system is used mainly when double insulation is possible, and no metal parts are accessible to the user.

insulating resistance the value of RESISTANCE between points that are insulated. The value can be used to calculate leakage current value, but it is important to remember that insulating resistance is nonlinear, that is, it can change with applied voltage, and that the value can be greatly affected by the presence of moisture.

insulation material with very high RESISTIVITY used for making INSULATORS.

insulator any material used to separate CONDUCTORS and prevent CURRENT flow between them. See INSULATION.

integrated circuit (IC) a CIRCUIT, also called a *microcircuit*, constructed entirely on a SILICON CHIP. The silicon is used to form TRANSISTORS, usually FETs; RESISTORS; and, less commonly, CAPACITORS.

This is made possible by using doped silicon as a conductor and silicon oxide as an insulator. The main advantages in the use of ICs are reliability and low price. A conventional circuit uses a number of components that have to be electrically connected. In such conventional circuits, the reliability decreases greatly as the number of components is increased, partly because each component becomes another potential source of failure, and partly because connections are unreliable.

The IC is manufactured as a single COMPONENT which keeps down the cost, and has the reliability of a single semiconductor component, despite the fact that it may carry out the actions of a circuit with thousands of components and connections.

integrated-injection logic a type of INTEGRATED CIRCUIT construction using BIPOLAR TRANSISTORS.

integrator a CIRCUIT that smooths out a PULSE into a half-wave of longer duration and lower amplitude. The application of repeated pulses to an integrator will have the effect of creating a steadily increasing voltage level. The term arises from the mathematical operation of integration. The output wave from an integrator is represented by an expression that is the time integral of the expression for the input signal. See Fig. 39.

intelligibility a measure of the utility of a communication system derived from the percentage of correctly received meaningless words in a sample transmission.

intensity 1. an alternative term for strength of an electric or mag-

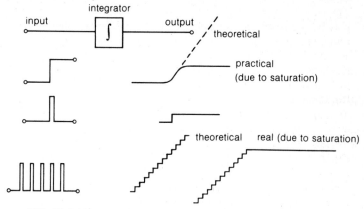

FIG. 39. **Integrator.** An integrator in symbolic form, and the effect of an integrator on typical circuit waveforms.

netic field (see FIELD-STRENGTH METER). **2.** the brightness of the SPOT (1) on a CATHODE-RAY TUBE.

intensity modulation the variation of brightness of the light SPOT (1) on the face of a CATHODE-RAY TUBE. Intensity modulation is essential to TV reception and is an option in most oscilloscopes.

intercarrier sound a system of sound IF used in COLOR TELEVI-SION receivers. The sound signal is frequency-modulated and transmitted on a frequency that is 6MHz higher than the video signal carrier. The RF and IF stages of a receiver deal with the entire band of frequencies. At the video DEMODULATOR the non-linear action of the diode mixes the sound and vision carriers, at their IF frequencies, to produce a 6MHz intercarrier frequency. This signal is still frequency-modulated with the sound signal and is filtered off to be amplified further and then frequency-demodu-lated.

interchangeability the ability to replace COMPONENTS in a CIR-CUIT without damaging the action of the circuit.

interconnection the making of CURRENT paths between circuits or devices, for example, between circuit boards in a SYSTEM.

interelectrode capacitance the CAPACITANCE between the sep-arate electrodes of a device. Capacitance of this type, such as the capacitance between the BASE and COLLECTOR of a transistor, may harm operation. For example, the capacitance will affect the phase of high-frequency signals and will also have to be charged and discharged when pulse signals are being used.

interface a CIRCUIT that allows SIGNALS to be passed between two otherwise incompatible systems.

interference any form of unwanted signal, which may be natural or man-made NOISE or a transmitted signal on the same frequency or on an adjacent frequency.

interlace the deliberate interweaving of lines on a TV RASTER.

interlock a form of safety SWITCH that allows power to be applied only when all necessary precautions have been taken, for example, when all cabinet doors are shut. In ROBOTICS interlocks are used to ensure that a robot cannot move if a human is in the operating area.

intermediate frequency (IF) the FREQUENCY used for the main amplifying stages in a SUPERHETERODYNE RECEIVER. The term arose because the frequency is intermediate between audio or video frequency and the incoming carrier frequency.

intermodulation the unwanted MODULATION of one frequency by another, caused by a NONLINEAR stage.

internal resistance the RESISTANCE of the internal circuits of a BATTERY or GENERATOR. Any source of signal or DC can be represented as a voltage supply in series with an internal resistance. The size of the internal resistance is found by measuring the change in output voltage per change in supplied current.

interrogating signal a PULSE signal that triggers a coded reply from a matching device, the TRANSPONDER. The system is used to identify friendly aircraft or aircraft that belong to a designated group. See also IFF, SECONDARY RADAR.

interrupt a PULSE signal applied to a microprocessor at an interrupt INPUT. This signal will cause normal processing to be suspended while a special software routine, the interrupt service routine, is run.

interrupter a DEVICE that switches a CURRENT on and off at set intervals.

interstage coupling a method of passing SIGNAL from one part of a circuit to another, or one stage to another.

intrinsic 1. of or relating to a specific quality of a material. 2. (of a SEMICONDUCTOR) consisting of pure undoped material.

intrinsic stand-off ratio the ratio of triggering voltage to applied voltage in a UNIJUNCTION.

inverse gain the GAIN figure for a TRANSISTOR operated with COLLECTOR and EMITTER (1) reversed.

inversion the action of an inverting AMPLIFIER on a SIGNAL. High and low voltages in the parts of the signal are interchanged, producing a WAVEFORM that is the mirror image of the INPUT.

inverter 1. an ANALOG or DIGITAL CIRCUIT or device that inverts a signal. The digital inverter is also called *NOT gate*. The digital inverter converts a 0 to a 1, and a 1 to a 0. 2. a circuit—*power oscillator* or *power converter*—that converts DC into AC.

inverting amplifier an AMPLIFIER whose OUTPUT is inverted with respect to the INPUT. Any single-stage transistor amplifier with input to the base and output from the collector load is an inverting amplifier.

ion the particle with CHARGE that is the result of an electron being added to or taken from a neutral atom. For a few materials, two electrons can sometimes be added or removed. See POSITIVE ION, NEGATIVE ION.

ion burn or **ion spot** a form of damage to the PHOSPHOR screen of a CATHODE-RAY TUBE. The damage is caused by NEGATIVE ION bombardment. It results in deactivation of the phosphor, so that it no longer emits light when struck by electrons. The negative ions result from remaining gas atoms that have been struck by electrons. See also ION TRAP, SCREEN BURN.

ionic conduction a form of CONDUCTION by the movement of positive or negative ions.

ion implantation a method of DOPING a semiconductor by bombarding the surface with IONS of the doping material. The technique can be controlled more closely than other doping methods and is particularly useful for making GALLIUM ARSENIDE FETs.

ionization the formation of IONS from neutral atoms. Radiation or high energy particles are needed to form ions in a gas at normal temperature. In the absence of radiation or such particles, a gas can be ionized by strong electric fields, particularly when the gas is at low pressure, or by temperatures of the order of several thousand kelvin. A completely ionized gas is called a *plasma*.

ionization chamber a space in which the production of IONS can be detected. An ionization chamber contains gas, which may be air, and two ELECTRODES with a POTENTIAL DIFFERENCE between them. If ions are produced in the gas, the movement of the ions between the electrodes produces a brief current that can be detected. This is the basis of most methods of measuring radiation from outer space or from radioactive materials. See also GEIGER-MÜLLER TUBE.

ionization current the current that flows in an IONIZATION CHAMBER as the result of ionization of the gas.

ionization gauge a method of measuring low gas pressures. A stream of electrons is used to ionize the gas within the gauge, and the IONS produced are attracted to a positive electrode. The ion current is related to the pressure over a wide range of low pressures.

ionizing radiation the particles or waves that will cause IONIZATION of any gas through which they travel.

ion loss the loss of energy because of HYSTERESIS in a magnetic core in an inductor. See also EDDY CURRENT, CORE.

ionosphere the outer layers of the atmosphere. These consist of air at low pressure that is ionized by the bombardment of particles from the sun. The IONIZATION is most intense when the sun shines directly on the layer, and the lower regions revert to a normal, less ionized state when the sun is shielded from them by Earth.

Several distinct layers have been identified. These shift in size and position according to the time of day, season, and condition of the sun. The ionosphere layers will reflect radio waves whose wavelength is more than 20 m or less than 6 mm. The layers were known at one time as the Appleton or Heaviside-Kennelly layers. See also C LAYER, D LAYER, E LAYER, F LAYER.

ionospheric focusing the increasing of signal strength at a RECEIVER because of the effect of the curvature of the IONOSPHERE layers. The curved layers act as a concave mirror for radio signals.

ion spot see ION BURN.

ion trap a method of preventing ION BURN in a CATHODE-RAY TUBE. The electron gun of a cathode-ray tube is tilted sideways, so any particle emerging straight along the axis of the gun will strike the neck of the tube. The ELECTRON BEAM is deflected back onto its correct path by a small magnet, but the heavier ions are not measurably affected and strike the tube neck.

Ion traps are used on monochrome TV tubes. Oscilloscope and color TV tubes are not affected greatly by ion problems.

irradiation the act or process of exposure of a material to IONIZING RADIATION or neutral particles.

ISO *abbreviation for* International Standardization Organization.

isolate disconnect from a supply, or make only an indirect connection.

isolating diode a DIODE used to allow PULSES to pass in one direction only.

isolating transformer a double-wound TRANSFORMER used to supply power to a system. Because of insulation between windings, there is no direct connection between the system and the external power source. Isolating transformers are extensively used in equipment repair shops to ensure that no equipment can be dangerous because of a direct connection to the AC power supply.

isolator a ferrite MICROWAVE device that allows a wave to travel along the waveguide in one direction, but absorbs any reverse wave.

isotropic (of a material) having measured quantities unaffected by direction. Many quantities, particularly magnetic quantities, are affected by the direction of metal CRYSTALS and thus are ANISO-TROPIC. That is, they have different values when measured along different crystal axes.

iterative impedance the value of IMPEDANCE measured in a NET-WORK where the impedance can be connected at one end of a network, thus making the network impedance at the other end identical. If a network has the same value of iterative impedance at both ends, that value is the CHARACTERISTIC IMPEDANCE. See also IMAGE IMPEDANCE.

J

jamming the deliberate interference with TRANSMISSIONS using signals on the same or a very close frequency.

JFET *abbreviation for* junction FIELD-EFFECT TRANSISTOR.

jitter an instability of DC or SIGNAL level resulting in small variations in amplitude or phase, particularly of a PULSE signal. Phase jitter on a pulse appears on an oscilloscope display as an irregular oscillation in the horizontal direction.

J-K flip-flop a FLIP-FLOP with two data inputs, labeled J and K. Its operation is summarized in the state table shown in Fig. 40b. If J and K are complements, Q takes the value of J at the next clock edge. If both J and K are low (0), the output does not change. If both J and K are high (1), the output Q will toggle, that is, change state after each clock edge. $\overline{Q}$ (NOT Q) is always the complement of Q.

Johnson noise see THERMAL NOISE.

Josephson effect the current flow in an insulating gap, the *Josephson junction* between superconductors (see SUPER-CONDUCTIVITY) at very low temperatures.

joule a unit of energy, equal to the watt-second. The kilowatt-hour unit is equivalent to 3.6 megajoules.

joule effect the heating effect of a CURRENT through a RESISTOR.

joule loss the loss of electrical energy in the form of heat when current flows through a resistor or any material with resistance.

jumper a connection, often temporary, between two points. A jumper connection is made by wires or plugs and is not part of a printed circuit.

junction 1. a contact between two materials. 2. the connection between differently doped pieces of semiconductor. 3. the contact between two metals in a thermocouple. 4. a point where wires are joined.

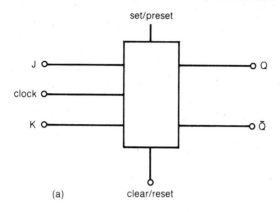

(a)

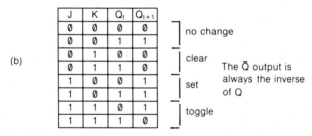

(b)

J	K	Q_t	Q_{t+1}	
0	0	0	0	no change
0	0	1	1	
0	1	0	0	clear
0	1	1	0	
1	0	0	1	set
1	0	1	1	
1	1	0	1	toggle
1	1	1	0	

The $\bar{Q}$ output is always the inverse of Q

FIG. 40. **J-K flip-flop.** (a) The J-K flip-flop symbol. (b) State table. It shows the state of the Q output before Qt and after Qt + 1, the clock pulse, and how this change is affected by the values of J and K.

junction box a box that contains circuit JUNCTIONS.
junction coupling a method of tapping a coaxial line into a cavity so as to pick up or inject MICROWAVE signals.
junction potential the ELECTRIC POTENTIAL between two materials that are in contact.
junction transistor a BIPOLAR TRANSISTOR employing two closely spaced junctions.

157

K

Karnaugh map a tabular method of designing a given logical function that enables the simplest implementation to be picked out quickly. The implementations of many logic functions using NOT, AND, and OR gates are not unique. For example, there are at least two ways to implement XOR. Unfortunately, Karnaugh mapping cannot be used for logical functions requiring more than four inputs except by computer-aided techniques.

K-band the MICROWAVE band covering approximately wavelengths of 8 to 27 mm.

keep-alive electrode a small discharge ELECTRODE in a gas-filled device that keeps enough IONS present to enable the main discharge to start quickly. See also TRANSMIT-RECEIVE SWITCH.

keeper a length of FERROMAGNETIC material placed across the poles of a permanent magnet. The purpose of the keeper is to complete the MAGNETIC CIRCUIT and so avoid the progressive demagnetization that occurs when a magnetic circuit is left open.

kelvin the unit of temperature based on the ABSOLUTE SCALE. Symbol: K. A kelvin is identical with a Celsius degree, but the zero of the Kelvin scale is ABSOLUTE ZERO. Named after Lord Kelvin (1824—1907), English physicist.

Kelvin effect see THERMOELECTRIC EFFECT.

Kelvin scale see ABSOLUTE SCALE.

Kennelly-Heaviside layer see E LAYER.

Kerr cell a method of controlling light by means of ELECTRIC FIELDS. Liquid in a Kerr cell will rotate the plane of polarization in POLARIZED light according to the size of the electric field applied across the liquid.

keystone distortion a form of picture shape DISTORTION. In optical projection, keystone distortion is caused by projecting a picture onto a screen that is not on the same axis. In early TV cameras, the effect was caused by the geometry of THE camera tubes, which used an off-center beam. Keystone distortion can also be caused if the amplitude of the HORIZONTAL TIMEBASE is affected by the amplitude of the VERTICAL TIMEBASE. See Fig. 41.

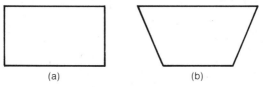

(a)   (b)

FIG. 41. **Keystone distortion.** (a) A rectangular shape. (b) The effect of keystone distortion on the shape in (a).

kilo- *combining form* denoting thousand. Symbol: k.

kilowatt-hour an energy unit used for home electric meters and equal to 3.6 MJ.

Kirchoff's laws the two laws describing the flow of CURRENTS and summation of VOLTAGES in steady state circuits. The first law states that the algebraic sum of currents at a point in a circuit must be zero—this is true only if no changing magnetic fields are present. The second law states that in any closed circuit, the algebraic sum of the ELECTROMOTIVE FORCES and of POTENTIAL DIFFERENCES must also be zero. See also SUPERPOSITION THEOREM.

Klystron *Trademark.* an ELECTRON-BEAM tube, formerly used in MICROWAVE transmitters and receivers. For low-power applications, such as local oscillators in microwave receivers, reflex Klystrons have been replaced by GUNN DIODES but the tubes are still used for higher power applications. The basis of the Klystron is an electron beam that passes through a cavity in which the electrons can be affected by resonant microwaves. Following this cavity, the BUNCHER the electrons move through the DRIFT SPACE in which no accelerating fields are applied. The movements of the electrons that were started by the fields in the buncher make the electrons gather into groups, or bunches, as they pass through the drift space. When the bunched electrons pass into another cavity, the catcher, they can give up their energy to the cavity. This energy will be considerably greater than the energy used in the buncher.

For very high power gains, Klystrons with several cavities can be used, with each cavity bunching the electrons tighter. The tendency for debunching, caused by electrostatic repulsion of electrons, can be counteracted by magnetic focusing. The REFLEX KLYSTRON is an oscillator that operates by reflecting the bunched electrons back to the buncher.

knife switch a form of SWITCH used now only for high voltage and high current supplies. The switch contacts are metal blades, which make a wiping contact with springy holders as the switch is closed.

L

labyrinth a type of LOUDSPEAKER enclosure. The sound wave from the rear of the loudspeaker cone is dispersed in a long, heavily DAMPED path.

ladder a network made by connecting alternating T-SECTIONS and pi sections of FILTERS.

lag 1. a signal or phase delay. 2. an image that is transmitted from a CAMERA TUBE for several frames after it has been removed from the face of the tube.

lagging current or **leading voltage** a CURRENT that lags in PHASE behind voltage.

lagging load a LOAD, such as an inductive load, in which the current PHASE lags behind the voltage phase. See also LEADING LOAD.

laminated core a TRANSFORMER core made by clamping together a set of thin metal laminations. The high resistance between laminations inhibits the flow of EDDY CURRENTS.

laser *acronym for* light amplification by stimulated emission of radiation, a generator of monochromatic coherent light (see COHERENT OSCILLATOR).

Normal sources emit PHOTONS of light in pockets, with the PHASE changing in each pocket of photons. This lack of phase stability is called *noncoherence.* It is impossible, for example, to set up any light interference experiment in which light of one source can interfere with light from a second source. This is because the phases of the waves are continually changing, canceling out any interference effects.

The laser generates light by allowing one group of waves to stimulate the emission of the next in a kind of RESONANT CAVITY. This ensures that every photon in a laser beam is synchronized and in phase. Laser light is termed *coherent light.* The synchronization can extend over a number of groups so large that waves separated by as much as a meter of space can still be in phase. Gas lasers use gas discharges as the source of light, and diode lasers use the LIGHT-EMITTING DIODE principle. Thus far the main applications for lasers in electronics are in optical transmission, using glass fibers. See also MASER.

latch the storage action of a FLIP-FLOP circuit. The output of a flip-flop is said to be latched if it remains constant despite changes in the input.

lattice filter a form of FILTER in which the impedances are arranged in a BRIDGE network.

L band the MICROWAVE band covering about 19 to 77 cm wavelength.

LCD see LIQUID CRYSTAL DISPLAY.

L-C network a RESONANT CIRCUIT containing capacitance and inductance. The tuned frequency is inversely proportional to the square root of the product of capacitance and inductance values.

lead 1. an electrical wire. 2. the advance in phase of one wave as compared with another.

lead-acid cell a form of secondary rechargeable cell, or ACCUMULATOR (2), constructed of lead plates in sulfuric acid. The lead plates are treated so that one acts as a CATHODE and the other as an ANODE.

leader the first part of a recording tape, which carries no coating. It is used only for threading and cannot be recorded on.

lead-in a connector through which an ANTENNA lead enters a building or a receiver.

leading current a CURRENT whose PHASE is earlier than the phase of VOLTAGE. The effect is also called *lagging voltage.*

leading edge a voltage change that is the start of a PULSE.

leading load a LOAD in which the PHASE of CURRENT is earlier than the phase of VOLTAGE such as a capacitive load. See also LAGGING LOAD.

leakage current the unwanted CURRENT caused by BREAKDOWN of insulation.

leakage flux the MAGNETIC FLUX generated by the primary winding of a TRANSFORMER but not coupled to or used by the secondary. Leakage flux is one source of loss in a transformer. See also HYSTERESIS, EDDY CURRENT.

leakage reactance the amount of REACTANCE in a TRANSFORMER that would correspond to the quantity of leakage flux.

lecher wires a method of measuring WAVELENGTH. A wave is coupled to a pair of parallel wires, and ANTINODES are found by using a small DIPOLE placed across the wires and connected to a lamp or a diode/meter arrangement. The distance between consecutive antinodes is equal to half a wavelength. A form of lecher wire can also be used for matching lines of different IMPEDANCES. See Fig. 42.

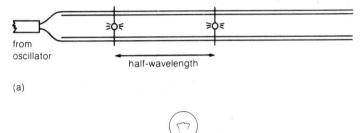

from oscillator

half-wavelength

(a)

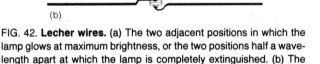

(b)

FIG. 42. **Lecher wires.** (a) The two adjacent positions in which the lamp glows at maximum brightness, or the two positions half a wavelength apart at which the lamp is completely extinguished. (b) The detector is a low-voltage flashlight bulb with two wires soldered to it.

LED see LIGHT-EMITTING DIODE.

left-hand rule see FLEMING'S RULE.

lengthened dipole a type of DIPOLE ANTENNA in which the rods have been moved apart to make the overall length greater than normal.

Lenz's law a law of electromagnetic induction. The induced current is in a direction such that it produces a MAGNETIC FLUX tending to oppose the original change of flux that causes it. For example, if the applied POTENTIAL DIFFERENCE across an inductor is increased, the increase of current through the inductor causes an increase of flux, and the voltage induced by this change of flux will be in a direction opposite to the applied potential difference, thus opposing the change. The practical effect is to reduce the rate at which such a change can be made.

LF see LOW FREQUENCY.

life test a timed use of COMPONENTS in a circuit to determine the time before failure. For reliable components such as semiconductors, an ACCELERATED LIFE TEST may have to be used.

lifetime the average time for which a current CARRIER exists in a SEMICONDUCTOR. This is the average time between electron-hole separation and recombination.

light-emitting diode (LED) a DIODE that emits light while CURRENT passes. The diode materials are compounds such as gallium arsenide or antimony phosphide. The material around the junction must be transparent so that the light can escape. LEDs are used as voltage indicators in electronic equipment. They are also used in ALPHANUMERIC displays for AC powered equipment.

The amount of current needed rules out the use of LEDs for most battery-operated applications, though at one time they were used for calculators and watches. For these low-energy applications, LIQUID-CRYSTAL DISPLAYS are now preferred.

light guide an application of FIBER OPTICS in which the glass fiber enables a light at one place to be seen elsewhere. For example, in a meter a fixed light source can be placed at one end of a fiber. The other end of the fiber can then be used to illuminate the scale needle, moving with the needle.

lightning rod a pointed spike facing upward on the top of a building. Under conditions of high CHARGE the air around the point ionizes and allows a quiet controlled discharge to take place, averting a lightning strike.

light pen a miniature PHOTOCELL mounted at the end of a penholder. The pen is used along with a computer program to write on the computer monitor screen, or select items by touching the pen to the correct place on the screen.

limb a part of a transformer core on which a winding is centered.

limiter a CIRCUIT that prevents the AMPLITUDE of a signal from exceeding a preset level. See also CLIPPER, CLAMPING DIODE.

line 1. see TRANSMISSION LINE. **2.** a horizontal scan of a TV screen.

linear of, in, along, or relating to a straight line. The term is generally applied to a relationship that produces a straight line on a graph. A linear SYSTEM is one in which the input signal V_{in} is related to the output signal V_{out} by an equation of the form

$$V_{out} = A \times V_{in} + B$$

where A and B are constants. Compare NONLINEAR.

linear amplifier an AMPLIFIER with low DISTORTION. The graph of output plotted against input for such an amplifier is nearly a straight line. For the region of the graph that is a straight line, the distortion will be minimal.

linear circuit a CIRCUIT for which the output/input graph is a straight line.

linear detector a DEMODULATOR for which the demodulated output is linearly proportional to the amount of modulation of the carrier.

linear network a NETWORK in which the graph of voltage plotted against current is always a straight line.

linear scan a SCAN across the face of a CATHODE-RAY TUBE at constant speed. This is not necessarily produced by a LINEAR TIMEBASE because the shape of the tube face and the characteristics of the deflection system can cause nonlinearity.

linear timebase a timebase whose waveform is a straight line SAWTOOTH.

line communication the sending of SIGNALS by wire as distinct from using radiated waves.

line flyback the part of a TV line TIMEBASE waveform in which the voltage or current returns to its starting value. See also RASTER.

line frequency 1. the FREQUENCY of horizontal scan in TV. In the US the term is also used to mean the mains supply frequency of 60Hz. **2.** the AC supply frequency of 60 Hz.

line of flux a line in a DIAGRAM that shows the direction of magnetic FLUX DENSITY. See also LINE OF FORCE.

line of force a term sometimes used to denote ELECTRIC FIELD direction, with LINE OF FLUX used to denote MAGNETIC FIELD direction. The two are more often used interchangeably.

line reflection the reflection of a signal from the far end of a

TRANSMISSION LINE. This will cause a STANDING-WAVE pattern if a steady signal is being sent continuously down the line.

line-sequential television an early type of COLOR TELEVISION SYSTEM in which the color signals for each line are sent in line sequence. The system is still used for specialized purposes, particularly narrow bandwidth transmissions.

line voltage 1. the VOLTAGE between the wires of a TRANSMISSION LINE. **2.** the AC supply voltage of 115 volts.

link 1. a system of TRANSMITTERS and RECEIVERS that connects two points by means of radio signals. **2.** a mechanical connection between SWITCHES that results in GANGING the switches. **3.** a wire connection between points on a circuit board.

lin-log response a specialized AMPLIFIER circuit characteristic. The circuit gives a linear response for small signals, and a logarithmic response for larger signals. This enables a large signal amplitude range to be used without overloading, a feature that is particularly important for radar receivers.

lip microphone a form of MICROPHONE case construction for outdoor use, designed to be used close to the lips, often as part of a headphone/microphone assembly. The benefit is that wind noise and other external noises are not heard as readily, since the sensitivity does not need to be high.

lip-sync a close synchronization of sound and picture. This is difficult to attain in film production, since sound is separately recorded, but not at all diificult in TV, since sound is recorded along with the picture. When true lip-sync is achieved in a film, the lip movement of speakers corresponds perfectly with the sounds.

liquid crystal a liquid whose long molecules can be aligned like the atoms of a CRYSTAL under the influence of an electric field. Alignment of these molecules causes light transmitted to the crystal to be POLARIZED. Many types of liquid-crystal materials are based on cholesterol, a fatty acid.

liquid-crystal display (LCD) a form of ALPHANUMERIC display that uses a POLARIZED filter fitted over a liquid crystal CELL or set of cells, and a reflector. With no FIELD applied to any cell, light passes through the filter and is reflected back from the rear of each cell. When an electric field is applied to a cell, the change in polarization of the cell makes it opaque, and with no reflected light the cell looks black from the outside.

The cells are usually arranged in the form of a SEVEN-SEGMENT DISPLAY but can take the form of words, diagrams, or any other shape required.

Lissajous figure a pattern that is traced out by two objects vibrating in directions at right angles to each other. The term is used in electronics to describe a technique that can be used for precise frequency comparison. One FREQUENCY a standard, is applied to one pair of deflection plates (*amplifier*) of an OSCILLOSCOPE. An unknown frequency is then applied to the other plates. If the signals are identical in amplitude and frequency, the resulting shape will be a straight line (frequencies in phase), a circle (90 degree phase difference), or an ellipse if the phases are of a different value. If the frequencies are not equal, the ellipse will rotate. For frequencies that are harmonically related (twice, three times, etc.), the display shows patterns in which a count of the peaks gives the ratio of frequencies. The slightest change of phase will cause a rotation of the pattern.

live (of an electrical device) at high voltage or with signal present.

load a COMPONENT or DEVICE that dissipates power from a CIRCUIT. The output of most electronic circuits is delivered to a load such as a heater, motor, antenna, loudspeaker, etc. All such loads require power to be converted, so the circuits that supply loads need to be capable of delivering the required power.

load impedance the IMPEDANCE of a LOAD as it appears to the OUTPUT circuit. Few loads are purely resistive, and many loads have MOTIONAL IMPEDANCE that will cause their apparent impedance value to vary with frequency or with time.

loading coils the INDUCTORS added to a CIRCUIT. For a TRANSMISSION LINE loading coils are connected at intervals to correct the value of CHARACTERISTIC IMPEDANCE. For the oscillator of a SUPERHETERODYNE RECEIVER loading coils are added in order to change the frequency band to a lower band, typically from medium wave to long wave.

load line a line drawn on a graph to represent the behavior of a LOAD. The usual load line is drawn on a graph of output current against output voltage. The line then represents the conductance of the load and enables the signal output voltage to be read off against signal current values. See Fig. 43.

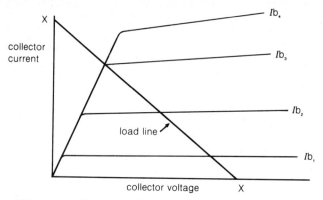

FIG. 43. **Load line.** A load line drawn on a current-voltage CHARAC-TERISTIC. The slope of the load line is proportional to the resistance of the load. In this example, the points on the load line are the values of collector current and voltage that can be obtained with this load value.

load matching the adjusting of a LOAD impedance to maximize transfer of power.

lobe the elliptical shape that is the contour line of constant field strength from a transmitting ANTENNA. The term is also applied to a receiving antenna to show the contour line of equal values of sensitivity. The radiation pattern of an antenna shows lobes of different field values. See Fig. 44.

lobe switching an electronic method of altering the direction of peak sensitivity of an ANTENNA by using several driven elements and varying the phase of signals to each.

local oscillator the oscillator within a SUPERHETERODYNE RE-CEIVER. The signal from the local oscillator beats with the incoming signal to form the INTERMEDIATE FREQUENCY.

lockin synchronize the FREQUENCY of an OSCILLATOR. One oscillator is said to be lockedin to another if it oscillates with the same frequency and, usually, with the same phase.

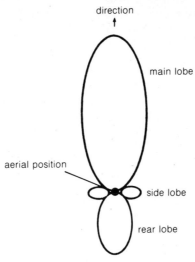

direction

main lobe

aerial position

side lobe

rear lobe

FIG. 44. **Lobe.** See this entry.

locking on 1. the establishing of automatic control over a varying quantity such as range, direction, or frequency. 2. an AUTOMATIC FREQUENCY CONTROL circuit that will lockon when manual tuning is almost correct, and from then on will maintain that tuning automatically. 3. the action of mechanical tracking systems, such as RADAR antennas, that lock onto a target and from then on follow it as long as it is within range.

locking relay a RELAY that can be mechanically or electrically locked in one position. The relay may be latched, meaning that the current to the coil can be supplied through relay contacts. Once the relay has been operated by another circuit, by means of a push button, for example, it will remain energized until power is removed.

logarithmic decrement the logarithm of the ratio of the peak amplitudes of two successive OSCILLATIONS of a system, a measure of the damping (see DAMPED) of oscillation.

logarithmic potentiometer a type of POTENTIOMETER in which the division ratio follows a logarithmic rather than a linear law.

LOGIC

logic a method of deducing a result (OUTPUT) from data (INPUT). George Boole showed in 1832 that all human logical processes could be described in terms of four simple steps, AND, OR, XOR, and NOT. See LOGIC GATE.

logic circuit a DIGITAL circuit that compares INPUTS and gives an OUTPUT according to a logic rule or set of rules. The elementary logic actions are carried out by LOGIC GATES whose actions are described in TRUTH TABLES or STATE tables.

logic diagram a DIAGRAM for a LOGIC CIRCUIT that shows LOGIC GATES in symbol form. This is not the same as an electronic circuit diagram, because it ignores power supplies, internal circuitry of gates, and any connections that are not logic signal connections. The positions of gates in the diagram need not correspond to their physical positions on a circuit diagram.

logic element a single LOGIC GATE or FLIP-FLOP.

logic gate a CIRCUIT that carries out a single LOGIC action. The OUTPUT of a logic gate is determined completely by the combination of INPUTS to that gate at that instant. The fundamental actions are AND, OR, XOR, and NOT, but the main logic gates manufactured in integrated circuit form are NAND GATE and NOR GATE. This is because any other gate action can be obtained by appropriate connections of NAND gates, or by different connections of NOR gates. Fig. 45 shows truth tables for NAND and NOR gates and illustrates how some other gate actions can be obtained by the use of NAND gates.

logic level the VOLTAGE range that is represented by a BINARY NUMBER 0 or 1.

logic one the VOLTAGE that corresponds to the BINARY NUMBER 1. For transistor-transistor logic circuits, this can be any voltage between 3.5 and 5.0V.

logic probe an instrument for reading LOGIC LEVELS. A flying lead is grounded, and the probe tip is placed on some connection in the logic circuit. The probe will then indicate whether the logic voltage is 0, 1, or an oscillating voltage.

logic symbol a shape on a DIAGRAM that represents a fundamental LOGIC action. The shapes that originated in the US have been used by manufacturers of integrated circuits and most users for a considerable time. See Fig. 46.

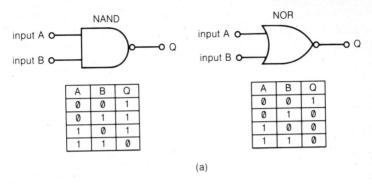

(a)

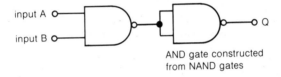

AND gate constructed
from NAND gates

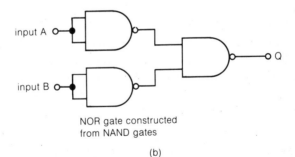

NOR gate constructed
from NAND gates

(b)

FIG. 45. **Logic gate.** The figure shows the symbols and truth tables
(a) and how the action of other gates (b) can be obtained by using
just one gate type, in this case the NAND gate.

LOGIC ZERO

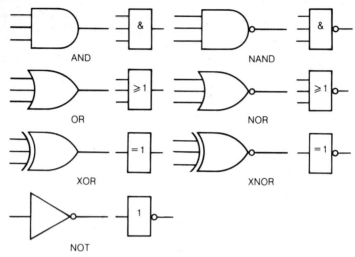

FIG. 46. **Logic symbols.** The symbols for gate actions. Two sets of symbols are shown, the usual international symbol on the left side of each pair, and the British Standard symbols on the right.

logic zero the binary 0 level of voltage. This is usually any voltage in the range 0 to +0.8V. See LOGIC LEVEL.

long-line pull a change of oscillator FREQUENCY when its output is fed to a long TRANSMISSION LINE.

long-persistence screen a form of PHOSPHOR screen once used for RADAR displays. This type of screen gives a bright glow when the ELECTRON BEAM strikes a SPOT and the glow continues at lower brightness, often in another color, for a considerable time afterward. This enables changes in the position of a radar target to be seen as a trail, gradually fading at the end. The tube must be viewed in darkness, and the operator has no control over persistence, so the tube cannot easily be erased.

At one time, the direct-view STORAGE TUBE was seen as a replacement for the long-persistence tube, but these were replaced in turn by computer-operated systems in which the information is retained in a computer memory, and the displays are TV type displays. This enables the displays to be viewed in more normal illumination, the persistence time to be controlled, and the screen to be cleared if the trails become confusing.

long-tailed pair see DIFFERENTIAL AMPLIFIER.

long wave the range of WAVELENGTHS that is measured in kilometers rather than in meters.

loop a connection of the output of a circuit or SYSTEM to its input, thus causing FEEDBACK.

loop antenna an ANTENNA in the form of a flat COIL. See also FRAME ANTENNA.

loop direction-finding see DIRECTION-FINDING.

loose coupling or **undercoupling** a form of tuned COUPLING in which the COUPLING COEFFICIENT has a low value, so only a small fraction of the signal in one circuit is transferred to the other. A loose coupling has two advantages: the BANDWIDTH can be small, and the connection of a load to the output does not greatly affect either the tuning or the amount of signal transferred. Compare TIGHT COUPLING. See also OVERCOUPLING, CRITICAL COUPLING.

Lorentz force the force on an ELECTRON or ION that is moving in a MAGNETIC FIELD.

loss a dissipation of power.

loss angle a measurement of LOSS from a CAPACITOR. A perfect capacitor would have a phase angle of 90 degrees, with current leading voltage. In practice, the angle is slightly less, as little as a thousandth of a degree, because of the loss of energy caused by slight electric HYSTERESIS. The difference is the loss angle.

lossless line a hypothetical TRANSMISSION LINE with no loss. A real line can be imagined as a length of theoretically perfect lossless line with a series resistor in which all loss of signal takes place.

lossy (of a material) dissipating energy, a term applied, for example, to an insulator that has current leakage.

lossy line a TRANSMISSION LINE that dissipates signal.

loudspeaker a TRANSDUCER that converts electrical signal energy into acoustic energy. Sometimes shortened to *speaker*. See also DRIVER, VOICE COIL, SPIDER, ELECTROSTATIC LOUDSPEAKER.

lower sideband the FREQUENCY RANGE obtained by subtracting each modulating frequency from the CARRIER FREQUENCY.

low frequency (LF) or **long-wave band** the band that covers the frequencies 30 to 300 kHz.

low-frequency compensation the AMPLITUDE and phase correction of a low-frequency signal (see PHASE-CORRECTION CIRCUIT). A signal containing low frequencies will have these frequencies attenuated and phase shifted by the coupling time constants of an amplifier. The low-frequency compensating circuits are designed to restore normal amplitude and phase.

low-level modulation the MODULATION of a CARRIER while it is at a low AMPLITUDE. The modulated carrier is then amplified. Amplitude modulation is usually carried out at high level, and frequency modulation at low level (on the oscillator). The use of low-level modulation means that subsequent radio frequency amplification must use fairly LINEAR stages.

low logic level the level 0 voltage, usually in the range of 0 to 0.8 V.

low-pass filter a FILTER whose PASSBAND extends from DC up to the cutoff frequency of the filter.

LSI *abbreviation for* large-scale integration, of the order of several thousand active DEVICES per CHIP. See also MSI, VLSI.

luminance the part of a VIDEO signal that carries light/shade information, the monochrome signal. See also COLOR TELEVISION.

luminance amplifier the AMPLIFIER in a COLOR RECEIVER that deals with the LUMINANCE signal.

luminescence the illumination of a material because of bombardment with electrons or other radiation.

lumped parameter a spread-out value that is treated as a single COMPONENT. A TRANSMISSION LINE for example, will have a value of inductance per meter and capacitance per meter. A length of line can therefore be simulated by using inductors and capacitors of the appropriate values, which are the lumped parameters.

M

machine code a PROGRAM in BINARY CODE for a MICROPROCESSOR.

magamp see MAGNETIC AMPLIFIER.

magic-T see HYBRID JUNCTION.

magnet see PERMANENT MAGNET, ELECTROMAGNET.

magnetic amplifier (magamp) a form of AC power AMPLIFIER with high power GAIN. A control winding on a MAGNETIC CORE (1) is used to control the magnetic characteristics of the core, and thus vary the REACTANCE of another winding. The power supply to the magnetic amplifier is AC, and the signals are DC or slowly changing signals.

magnetic circuit a closed path in a magnetic material. The magnetic circuit conducts MAGNETIC FLUX in the same way that a metal circuit conducts current.

magnetic core 1. a piece of magnetic material on which an inductor can be wound. 2. in computing, the miniature magnetic rings formerly used for constructing core storage memory in large computers. The principle is that a binary 1 is stored in the form of one direction of magnetic flux, and a binary 0 in the form of the opposite direction. The cores are read and written by passing current through wires that thread through each core.

Modern computers almost always use SEMICONDUCTOR memory.

See also COINCIDENT HALF-CURRENT SWITCHING.

magnetic damping a method of damping (see DAMPED) mechanical OSCILLATIONS by using induced currents. The oscillating object is attached to a piece of metal that moves between the poles of a magnet. The EDDY CURRENTS induced in the metal set up forces that oppose the motion (LENZ'S LAW), thus damping the motion.

MAGNETIC DEFLECTION

The advantage of magnetic damping is that the damping force is proportional to speed. This means that the maximum damping is achieved when the movement is a maximum, and there is practically no damping when the object is almost at rest.

magnetic deflection the DEFLECTION of an ELECTRON BEAM by a MAGNETIC FIELD.

magnetic dependent resistor any resistive material, mainly SEMICONDUCTORS, whose RESISTANCE value is altered by the presence of a MAGNETIC FIELD. See MAGNETORESISTANCE.

magnetic disk a plastic DISK coated with magnetic powder that is used as an external storage medium for computers. The storage method is based on direction of magnetism (see MAGNETIC CORES) and the disk is spun while a READ/WRITE HEAD is put into contact or near-contact with its surface. This affords quick access to any part of the disk.

magnetic field the space around any form of magnet in which force effects on FERROMAGNETIC materials can be detected.

magnetic flux a flow, or FLUX (1), of magnetism. This is most easily measured in terms of the ELECTROMOTIVE FORCE generated when a circuit surrounding the flux is removed from the flux.

magnetic flux density a measure of the strength of MAGNETIC FIELD in a material. Symbol: B; unit: weber per square meter. See also FLUX DENSITY.

magnetic focusing a method of focusing an ELECTRON BEAM by using MAGNETIC FIELDS.

magnetic hysteresis the NONLINEAR relationship between FLUX DENSITY in a material and the MAGNETIC FIELD that produces it.

In Fig. 47, the initial part of the graph that starts at the origin is traced out when a magnetic field is applied to a material that is not initially magnetized. Increasing field strength leads to the condition of SATURATION at which increases of field strength cause only comparatively small changes in flux density.

If the field is then reduced, the shape of the graph from the saturation point downward is not the same as the shape when the field was being increased. There will be flux density remaining in the material even when the external field is zero, and this value of flux density is called the REMANENCE.

If the field is reversed and its strength increased, it will oppose

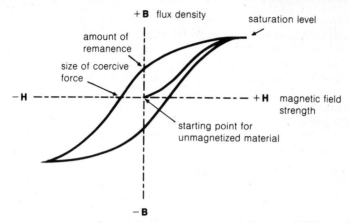

FIG. 47. **Magnetic hysteresis.** The curve shows the once-only part of the trace for an unmagnetized material, and the saturation, remanence, and coercive force points on the graph. The graph is normally plotted with units for **B** that are much larger than the units for **H** in order to make the graph of reasonable size.

the flux density direction, and the flux density will decrease. The field value needed to reduce the flux density to zero is called the COERCIVITY and is a measure of the permanence of the magnetism.

Increasing the field now reverses the direction of the flux density until saturation is reached in the opposite direction. Reduction of field strength from this point causes another path to be traced, making the trace symmetrical. The area enclosed by the path of this HYSTERESIS LOOP represents the amount of energy lost in the whole cycle. Note that in hysteresis diagrams the units are always chosen so as to produce diagrams of reasonable scale.

magnetic leakage a loss of FLUX (1) from a MAGNETIC CIRCUIT. Unlike current in a metallic circuit, flux is not totally confined to magnetic material, and the amount of flux that strays from the intended path is the LEAKAGE FLUX.

magnetic lens a lens for focusing an electron beam, produced by using magnets to form a magnetic field that will focus the beam. The magnetic lens may use permanent magnets, but it is more common to use electromagnets to form an ELECTROMAGNETIC LENS, whose focus can easily be adjusted by varying the current in the electromagnet(s).

magnetic memory a MEMORY system in which data are represented as direction or intensity of magnetization of a material. The advantage of such a system is that it is NONVOLATILE, that is the data are not erased when power is removed from the system.

magnetic moment a quantity sometimes used to measure the strength of a magnet. The torque acting on a bar magnet in a field is the vector (cross) product of magnetic moment and the FLUX DENSITY of the surrounding field.

magnetic pickup a type of phonograph PICKUP that makes use of electromagnetic induction. See MOVING-IRON PICKUP, MOVING-COIL PICKUP.

magnetic pole a point around which magnetism is concentrated. Any magnet that has an air gap will have a pole on each side of the air gap.

magnetic recording a method of recording electrical signals as variations of the MAGNETISM of a material. For audio recording, the material is a finely powdered iron oxide or chromium oxide, or a metal, usually iron, coated on plastic tape. The tape is fed past a recording head at a steady speed. The head consists of a FLUX (1) path that is almost totally closed, but with a tiny gap that touches the tape.

This causes an intense FIELD to cross the tape and thus magnetize the particles on the tape. The signal to the tape head takes the form of audio current signals on a winding around the magnetic material. On playback, the tape is again drawn at the same constant speed past a similar head, often the same head. This time, the variations in magnetization of the tape cause a varying flux in the magnetic circuit of the head and thus induce varying voltages in the winding. See also MAGNETIC TAPE, PAPER TAPE.

Because the graph of tape magnetism plotted against head coil current is a nonlinear (HYSTERESIS) shape, some form of BIAS is needed to achieve acceptable results. This is achieved by using high-frequency signals. See also AZIMUTH ANGLE.

magnetic saturation the point at which the magnetic FLUX DENSITY of a material ceases to increase perceptibly when the magnetizing force is increased. At this point, the PERMEABILITY of the material is virtually the same as that of the air around the material.

magnetic screening the protecting of COMPONENTS against the effect of MAGNETIC FIELDS. This is necessary if unwanted magnetic fields are present near CATHODE-RAY TUBES, TRANSISTORS, INTEGRATED CIRCUITS or INDUCTORS. The magnetic screening is a closed box made of highly permeable material so that flux passes through the screening material in preference to any other path.

magnetic shunt an alternative path for MAGNETIC FLUX consisting of a piece of high-permeability material whose position on a magnetic path can be varied so as to control the amount of flux in another path.

magnetic tape a plastic tape coated with magnetic material for the purpose of recording. Iron oxide is the most usual coating, but chromium (IV) oxide and metallic iron have also been used for high-quality recording. See MAGNETIC RECORDING, PAPER TAPE.

magnetic tuning a tuning method for MICROWAVE cavities. Part of the RESONANT CAVITY is made of FERRITE whose PERMEABILITY can be altered by the steady field from an electromagnet.

magnetism a field effect that produces strong forces on ferromagnetic and paramagnetic materials. MAGNETIC FIELDS are produced by circulating currents. These can be currents in wire conductors, or the atomic-scale circulation of electrons.

A number of materials are affected to some extent by magnetism. *Paramagnetic* materials will attempt to align their magnetic axes with the field, and *diamagnetic* materials will attempt to place their magnetic axes at right angles to the field. FERROMAGNETIC materials are a form of paramagnetic material, for example, iron, on which strong forces act. Many ferromagnetic materials can be permanently magnetized.

magnetization the effect of creating a magnetic FLUX DENSITY in the space around a material. Magnetization may be due to the presence of a PERMANENT MAGNET or to the MAGNETOMOTIVE FORCE caused by electric current.

magnetize make a material permanently or temporarily magnetic.

magnetomotive force a cause of MAGNETIZATION. Symbol: *Fm;*

unit: AMPERE-TURN. The magnetomotive force of a coil causes MAGNETIC FLUX in a MAGNETIC CIRCUIT.

magnetoresistance a change of resistance caused by magnetism. See MAGNETIC DEPENDENT RESISTOR.

magnetostriction a stress caused in a material by MAGNETIZA-TION. The forces between atoms in a magnetized material cause stresses that can affect the dimensions of the material. *Positive magnetostriction* causes an increase in length of a material when it is magnetized. *Negative magnetostriction* causes a decrease in length of a material when it is magnetized. Alternating magnetic fields applied to either type of material will cause mechanical vibrations, which are used in ULTRASONIC generators. The familiar whistling sound from a TV receiver is caused by magnetostrictive vibrations of the core of the line-output transformer.

magnetron the earliest type of high-power MICROWAVE oscillator. The principle is based on the creation of pulsed beams of electrons that travel in circular paths in a vacuum. By making the ELEC-TRON BEAM of each electron travel across the mouth of one of several identically tuned cavities, the cavities are set in RESO-NANCE at a microwave frequency. The oscillation reaches a peak when the speed of circulation of the electrons is such that one cycle of the electrons around the beam path takes the same time as one cycle of the microwave frequency.

Stability of frequency is achieved by making connections (see STRAPPING) between alternate cavities. The microwave signal energy is tapped by a loop in one cavity that is coupled to an external WAVEGUIDE. See Fig. 48.

magnitude the ABSOLUTE VALUE or amplitude of a signal expressed as a positive value.

mainframe the largest type of computer.

majority carrier the charged electron or hole responsible for more than 50% of the current flow in a SEMICONDUCTOR. Compare MINORITY CARRIER.

major lobe the longest LOBE in an antenna radiation pattern. This corresponds to the direction of transmission or reception for a DIRECTIONAL ANTENNA.

make close a CIRCUIT.

make-and-break circuit a form of mechanical circuit INTER-

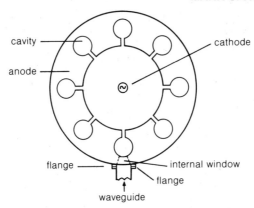

cavity

anode

cathode

flange

internal window

flange

waveguide

FIG. 48. **Magnetron.** A simplified cross section through a magnetron. The strapping between alternate cavities is not illustrated.

RUPTER that converts steady current into pulses. See also CHOPPER.

Manganin *Trademark.* a copper-manganese alloy that is used for constructing WIRE-WOUND RESISTORS. It is particularly suitable because of comparatively large RESISTIVITY low TEMPERATURE COEFFICIENT of resistivity, and small CONTACT potential with copper.

man-made noise the NOISE caused by man-made DEVICES, such as automobile ignition systems, thermostats, light switches, and fluorescent lights.

Marconi antenna a simple vertical wire ANTENNA.

mark the part of an on/off cycle of voltage that is the on-time. If a voltage is switched on for 2 ms and is off for 3 ms, then the mark is the 2 ms time for which the voltage is on. The off-time is called the *space.* The mark-space ratio in this case would be 2/3.

marker pulse a PULSE used as a time reference in PULSE-TIME MODULATION and MULTIPLEXER systems.

mark-space ratio the ratio of on-time, defined as time for which voltage is supplied, to off-time for a rectangular pulse. The mark-space ratio for a perfectly square pulse is 1:1. See Fig. 49. See also MARK, DUTY CYCLE.

MASER

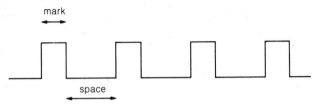

mark/space ratio $= \frac{1}{2}$

duty cycle $= \frac{1}{3}$

If mark/space ratio $= \frac{x}{y}$ then duty cycle $= \frac{x}{x+y}$

$$D = \frac{M}{M+1}$$

$$M = \frac{D}{1-D}$$

where M = mark/space ratio and D = duty cycle

FIG. 49. **Mark-space ratio.** The mark-space ratio for a pulse waveform, and how it is related to the duty cycle.

maser *acronym for* microwave amplification by stimulated emission of radiation, a form of MICROWAVE amplifier that uses the electron energy in a gas or a solid crystal. The electrons in the material are excited to a high energy level, and the incoming signal is used to trigger the return to normal level, thus releasing energy.

Solid crystal masers run at low temperatures (liquid helium temperatures of about 4 degrees K) and offer high power gain with low noise. Gas masers are used as precise microwave oscillators. Similar principles are used for LASERS.

mask shield areas of a CHIP in order to make a process such as ETCHING or DIFFUSION follow a desired shape pattern.

masked ROM a type of permanent ROM memory consisting of electrical connections. Each unit of the memory consists of a permanent connection to ground (logic 0) or to the positive supply terminal (logic 1), so that the data are fixed when the memory integrated circuit chip is manufactured by the process called *masking.*

This type of memory is NONVOLATILE because the retention of data does not depend on the power supply being maintained.

master oscillator a highly stable OSCILLATOR that is used as the source of a CARRIER signal. The master oscillator output is to a BUFFER stage or stages, to ensure that the effect of connecting a LOAD to the master oscillator is negligible.

master-slave flip-flop a form of double FLIP-FLOP in which the first unit, the master, drives the second, the slave. See J-K FLIP-FLOP.

matched termination a cable TERMINATION at which there is no signal REFLECTION. The termination is the load at the end of a TRANSMISSION LINE or WAVEGUIDE and if correctly matched it produces no STANDING WAVES in the line.

matching obtaining equality of the magnitude of LOAD and OUTPUT impedance. This ensures maximum power transfer in the case of a fixed output resistance (see MAXIMUM POWER THEOREM).

maximum power theorem a theorem for finding matching load value. For a given output impedance, the maximum power theorem shows that maximum power is developed in a load whose resistance is exactly equal to the output resistance, and whose reactance is exactly equal and opposite to the reactance of the output.

The converse is not true. If the load is fixed, then maximum power in the load is achieved when the output resistance of the driving stage is as low as possible.

Maxwell's equations the equations of electromagnetic waves that link magnetic field strength (H), electric displacement (D), magnetic flux density, (B) and electric field strength (E). The solution of the equations takes the form of a wave whose velocity depends on the PERMEABILITY and PERMITTIVITY of the medium in which the wave travels.

M-derived filter a FILTER that has a reactive component in one arm and a RESONANT CIRCUIT in the other. See Fig. 50.

mean time between failures (MTBF) the average length of time a COMPONENT will perform as predicted before breaking down. This may be measured under working conditions or calculated from ACCELERATED LIFE TESTS conducted under stressed conditions. See also FAILURE RATE, RELIABILITY.

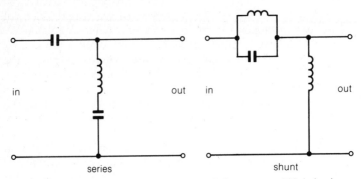

FIG. 50. **M-derived filter.** The series and shunt types of M-derived filter circuit.

mechanical resonance a forced mechanical OSCILLATION at RESONANT FREQUENCY. Mechanical resonance can cause failure of some mechanical parts, but it often can be useful, for example, in a device using a pendulum.

Mechanical resonance is also responsible for resonant peaks in MOTIONAL IMPEDANCE but this can be turned to advantage if a large amount of power must be converted to mechanical form at a certain frequency.

medium frequency (MF) the RF band that covers the range of about 300 to 3000 kHz.

medium wave (MW) the WAVELENGTH range of about 100 to 1000 m.

mega- *combining form* denoting one million. Symbol: M.

Megger *Trademark.* an instrument for measuring INSULATING RESISTANCE. The original trade name denoted an instrument that combined a handwound high-voltage generator with a sensitive current meter. Modern versions use DC inverters and electrometers.

memory a circuit for storing data in BINARY CODE particularly for a computer. This applies to storage as current in one of a pair of transistors (STATIC RAM), charge in a capacitor (DYNAMIC RAM), or connections in an integrated circuit (MASKED ROM). Memory systems can also be constructed using magnetic effects, such as MAGNETIC CORE stores, BUBBLE MEMORY, MAGNETIC TAPE, and MAGNETIC DISK.

Memories that depend on current or charge are generally volatile, because the data are lost when power is switched off. Magnetic and masked ROM memories are not volatile, because the stored data remain in memory until replaced by other data.

See also ROM, PROM.

memory location the binary or hexadecimal number ADDRESS required to gain access to a specific memory location.

mesa transistor a form of silicon junction transistor construction. The BASE of the transistor is formed on a portion that is raised above the level of the COLLECTOR and EMITTER like the steep-walled flat-topped hill called a mesa, a Spanish word for table.

mesh a NETWORK of CURRENT paths.

mesh current the CURRENT in a network MESH. It can be used to form equations of network action, using KIRCHOFF'S LAWS or NORTON'S THEOREM.

mesh-PDA tube a specialized form of POST-DEFLECTION ACCELERATION for INSTRUMENT TUBES. The DEFLECTION SENSITIVITY of an electrostatically deflected cathode-ray tube is very low if the accelerating voltage is high, of the order of 10 kV.

Modern oscilloscopes require tubes that have very high deflection sensitivity so that transistor deflection amplifiers can be used. At the same time the trace must be bright for high-speed traces to be visible, and this demands large accelerating potentials.

The mesh-PDA tube meets these requirements by placing a wire mesh of high transparency between the end of the DEFLECTION PLATES and the FINAL ANODE. The mesh separates the fields, so the region of the deflection plates can use low velocity electrons, which are accelerated after passing through the mesh.

metal film resistor a form of FILM RESISTOR that uses a thin metal film as its conductor.

metalized paper capacitor a form of PAPER CAPACITOR that uses a metalizing coating on the paper, rather than metal strips.

metalizing the process of coating a material with metal. This can be done chemically, by ELECTROPLATING (on metals), or by chemical reduction (on other surfaces). Another method is to evaporate the metal in a vacuum, as used for depositing thin films of aluminum. In SPUTTERING, a third method, the material to be coated is placed in a low-pressure gas environment in which a GAS DIS-

CHARGE is operating, with the metal forming the coating material being used as one electrode. Sputtering is especially suitable for depositing powdery films of metals.

Combinations of methods can be used, and it is fairly common to find an insulating material coated with a thin conducting film by sputtering, and then electroplated.

metal rectifier a RECTIFIER that makes use of a metal semiconductor CONTACT. Copper-oxide and selenium rectifiers were used to a considerable extent at one time.

meter 1. a measuring instrument whose reading is usually in the form of a needle positioned against a scale. **2.** the basic SI unit of length. Symbol: m. The length of the meter is specified in terms of the number of wavelengths of light from a specified source.

meter-protection diode a DIODE used in pairs wired across a meter movement. The principle is that the diodes form a high-resistance path for voltages up to and exceeding the normal voltage across the meter terminals. For a large overload voltage, however, one or both diodes will conduct, thus preventing damage to the meter. The protection diodes are particularly effective in preventing damage from transient pulses.

meter resistance the INTERNAL RESISTANCE of a METER movement, including any internal shunts.

MF see MEDIUM FREQUENCY.

mica a natural INSULATOR that can be split into thin sheets. The silver form is preferable to the ruby form for use with electronics.

mica capacitor a CAPACITOR made by metalizing mica plates. Mica capacitors have low losses, low TEMPERATURE COEFFICIENTS, and stable characteristics.

micro- *combining form* in electronics denoting one millionth. Symbol: μ.

microcircuit see INTEGRATED CIRCUIT.

microelectronics the design and construction of electronic devices using INTEGRATED CIRCUITS.

micron a millionth of a meter, that is, a *micrometer*.

microphone a sound to electrical wave TRANSDUCER. Many types of microphones have been devised. The first was the carbon-granule type used in early telephones. Most modern types are ELECTRODYNAMIC RIBBON MICROPHONES, MOVING-COIL MICROPHONES, MOVING-IRON MICROPHONES. CAPACITOR MICROPHONES which once seemed to be obsolescent, were revived when ELEC-TRETS became commercially available.

The main features of a microphone influencing choice are sensitivity, directionality, size, quality, and price. Sensitivity is measured in terms of peak-to-peak output for a given sound pressure wave and is usually low, particularly compared with the sensitivity of the human ear. Many microphone types are directional, and the way in which a microphone is housed can enhance the main lobe of sensitivity. Size varies considerably, but ribbon microphones tend to be larger than the other types.

The highest quality is obtained from ribbon microphones, mainly because the MECHANICAL RESONANCES can be at frequencies well above the audio range. Ribbon microphones are generally the most expensive, and capacitor microphones the cheapest. The enclosure in which the microphone is housed is at least as significant acoustically as the construction of the microphone itself.

microphony an unwanted electrical signal generated in response to sound waves. Some electronic components, particularly plates of air-spaced variable capacitors, can vibrate when struck by sound waves and cause MODULATION of electrical signals. Radio tubes were particularly prone to this problem, which seldom is encountered in transistorized receivers.

microprocessor a VLSI integrated circuit that carries out computing actions. The microprocessor is a PROGRAMMABLE assembly of REGISTERS and GATES. A master CLOCK determines the speed at which instructions are executed, and the program is in the form of number codes stored in consecutive locations in memory. Given the correct starting location in memory, the microprocessor can load in the instruction codes one by one and execute them.

The processes that can be carried out include simple arithmetic such as addition and subtraction of small numbers; LOGIC actions such as AND GATE, OR GATE, XOR GATE, and INVERTER (1); and SHIFTING and ROTATION. All the operations of a computer are obtained from these few logic actions by suitable programming.

microprogram the instructions built into a MICROPROCESSOR by the manufacturer. These instructions enable the microprocessor to carry out the fixed set of actions under external program instruction that are described in the manufacturer's specification.

microstrip a form of TRANSMISSION LINE consisting of a metal ribbon that acts as a ground plane and separated by a ribbon of insulator from a narrower ribbon of conductor, which is the signal conductor. The CHARACTERISTIC IMPEDANCE depends on the width, and is typically between 10 and 50 ohms.

microwave the range of radio WAVELENGTHS from approximately 35 cm down to fractions of a millimeter. These cannot easily be generated or amplified by the conventional electronic devices used for lower radio frequencies and are transmitted more readily through WAVEGUIDES than along conventional copper conductors.

microwave oven an electrically operated oven in which food is the target of a microwave beam from a MAGNETRON. The frequency is 2.45 GHz, because this corresponds to a resonance of hydrogen atoms in water. Any water present in food will therefore be rapidly heated internally. Objects with very low water content are not heated. Metal objects must not be present, because they can damage the magnetron by reflecting power back into the system.

microwave generator a vacuum or semiconductor device used as an OSCILLATOR at MICROWAVE frequencies. This includes KLYSTRONS, CARCINOTRONS, MAGNETRONS, GUNN DIODES and Josephson junctions (see JOSEPHSON EFFECT).

microwave tube any vacuum MICROWAVE device, including the KLYSTRON, TRAVELING-WAVE TUBE, MAGNETRON.

mil a unit of length equal to one thousandth of an inch, or 0.0254 mm.

Miller effect the effect of NEGATIVE FEEDBACK from output to input of an amplifying device through stray or internal CAPACITANCE. The capacitance is the total of interelectrode capacitance and stray capacitance. For inverting amplifiers, the Miller effect makes this capacitance appear to be multiplied by the voltage gain.

Miller integrator a wide-range INTEGRATOR circuit that depends for its action on a CAPACITANCE connected between the input and output of an INVERTING AMPLIFIER. See Fig. 51.

Miller timebase a TIMEBASE circuit that consists of a squarewave generator combined with a MILLER INTEGRATOR.

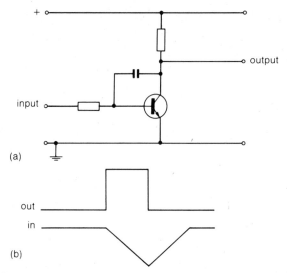

FIG. 51. **Miller integrator.** (a) The basic circuit and (b) typical waveforms.

milli- *combining form* in electronics, meaning a thousandth. Symbol: m.

millimetric waves the WAVELENGTHS in the range 1 to 9 mm.

MIL specification a specification drawn up for equipment intended for use by the US armed forces, either as original equipment or as spares.

minimization the application of LOGIC theory to a LOGIC GATE circuit. Minimization applies logic theory to find redundant gates, and often enables a logic action to be carried out simply. See KARNAUGH MAP.

minority carrier the electron or hole in a semiconductor that carries less than 50% of the current. Compare MAJORITY CARRIER.

mismatch the inequality of OUTPUT IMPEDANCE and LOAD IMPEDANCE. Mismatch can cause loss of power and REFLECTIONS along a line.

mixer the frequency changer stage of a SUPERHETERODYNE RE-

CEIVER. The mixer has two inputs, one at received signal frequency, which is modulated, and one from the local oscillator. The result of mixing is a signal at INTERMEDIATE FREQUENCY that carries the original modulation.

MKS system a system of units used in science and engineering, based on the meter, kilogram, and second as fundamental mechanical units, supplemented by the ampere for electrical work.

mnemonics in electronics, a set of simple words used to write programs for a MICROPROCESSOR.

mode 1. the FREQUENCY and WAVEFORM of a resonating system. 2. the classification of a WAVE in terms of its polarization (see POLARIZED sense 1). The three modes are TEM WAVES, TE WAVES and TM WAVES.

modem *acronym for mo*dulator-*dem*odulator, a device used to convert digital signals into a form suitable for long-distance transmission, and to reconvert received signals.

The modem is widely used for linking computers over telephone lines and via radio links, where permitted. The MODULATOR section translates the data bits of 0 and 1 into tone signals at frequencies representing 0 and 1. The data then are transmitted as a set of tones, using rates of transmission that are slow by computing standards. At the receiving end, the DEMODULATOR converts the received tones into data signals again. See also SERIAL TRANSMISSION.

modulation the alteration of a CARRIER (1) wave so that it bears information. The most elementary form of modulation, often regarded as not being a form of modulation, is Morse code, in which the carrier is switched on and off. For carrying audio or video signals, AMPLITUDE MODULATION or FREQUENCY MODULATION is used.

Any of these modulation methods will produce a signal that consists of a carrier and a set of SIDEBANDS in which the sidebands carry the information. The most common forms of modulation are wasteful of transmitted energy, because only a fraction of the total radiated power is in the form of the sidebands that carry the information.

Several systems exist that suppress one sideband and/or the carrier in order to use transmitter power more efficiently. These

SINGLE-SIDEBAND systems require special DEMODULATION circuits, and the carrier frequency must be considerably higher than the highest frequency that will be modulated onto it. For MICRO-WAVE transmissions, it is generally easier to pulse the carrier on and off and modulate the pulsing system. At the receiver, various types of demodulators are needed to recover the form of the original signal.

modulation factor see DEPTH OF MODULATION.

modulation index a figure used in calculating BANDWIDTH for frequency-modulated signals (see FREQUENCY MODULATION). The modulation index is obtained by dividing the maximum frequency change of the CARRIER by the maximum frequency of the modulated signal. The modulation index should be a small fraction.

modulator any CIRCUIT that modulates a CARRIER.

Moiré pattern an image formed by light passing through two or more grids. Sometimes called *strobing* by camera operators. Moiré patterns are a problem in TV pictures because of the use of scanning in lines and the vertical bar structure of the aperture grille. The effect is that pictures containing horizontal or vertical stripes appear to contain elaborate moving patterns, often colored.

monitor a VISUAL DISPLAY UNIT commonly used with computers.

monochromatic radiation radiation consisting of a single frequency or a small range of frequencies. An unmodulated CARRIER that is a perfect sine wave is a monochromatic radiation.

monochrome a picture or display using shades of one color only, often referred to as black-and-white even though computers make use of monitors whose screen color may be green or amber.

monolithic circuit a CIRCUIT made in one piece, that is, an INTE-GRATED CIRCUIT fabricated from a single piece of silicon.

monophonic (of a sound-reproducing system) using one sound channel only. Compare STEREOPHONIC.

monoscope a form of TV SIGNAL-GENERATOR tube used to provide a pattern such as a test chart. Simple dot and bar patterns can be generated by signal generators, but the monoscope uses camera-tube techniques to generate a more elaborate picture signal. Monoscopes are being replaced by computer-generated test patterns.

monostable or **one-shot** or **single-shot** a type of MULTIVIBRA-TOR with one stable state. A TRIGGER PULSE is used to place the monostable in its unstable state and, after a short time, the circuit switches back to its original state. The length of time is deter-mined by a capacitor-resistor TIME CONSTANT. The device is used to generate a pulse of known width from a trigger pulse. See Fig. 52.

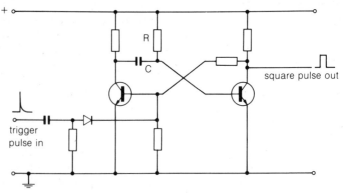

FIG. 52. **Monostable.** A typical monostable circuit. The duration of the pulse output is determined by the time constant **CR.**

Morse code an established system of communication. Morse code uses two signals, the short dot and the long dash, to code letters of the alphabet and digits. Morse can be used for communication along long lines, and for radio signaling by switching a CARRIER on and off.

The advantage of Morse is that it makes efficient use of a low-power carrier and is highly noise-immune. Today, it is used mainly for emergency systems, but its influence on radio practice still lingers on in the form of the requirement for radio amateurs to take a test in Morse.

MOS *acronym for* metal-oxide semiconductor. See FIELD-EFFECT TRANSISTOR.

mosaic a set of insulated conducting dots on an insulating surface. The mosaic was a feature of early TV camera tubes in which the dots acted as CAPACITORS storing CHARGE quantities that formed a charge image corresponding to the light image.

MOS capacitor a metal oxide semiconductor CAPACITOR formed on a chip by using doped (see DOPING) silicon for plates and silicon oxide as the INSULATOR. MOS capacitors are the basis of CHARGE-COUPLED DEVICES and of DYNAMIC RAM devices.

MOSFET see FIELD-EFFECT TRANSISTOR.

MOS IC an INTEGRATED CIRCUIT constructed using MOSFETS as the active components. See FIELD-EFFECT TRANSISTOR.

MOS transistor see FIELD-EFFECT TRANSISTOR.

motherboard a baseboard for a computer circuit used in small computers. The main parts of the computer are on a single board, called the motherboard. This board is fitted with sets of sockets, each of which can connect to another board. All the computer signals are made available at each socket, so the facilities of the computer can be greatly extended by plugging in extension boards.

The system is often referred to as having *open architecture* in the sense that the computer can be extended by using boards made by suppliers other than the original manufacturer. Computers of this design type, such as the Apple-2, can have long working lives, because it appears always possible to upgrade the design rather than replace the machine.

motional impedance a type of IMPEDANCE found in an electromechanical TRANSDUCER. A transducer such as a LOUD-SPEAKER converts electrical signals into mechanical motion. The current that a signal of constant voltage can pass will be greatest if mechanical motion is prevented, will be least if the motion is completely unimpeded. This means the electrical impedance will also vary with the resistance to mechanical motion, particularly if there is mechanical resonance. Compare BLOCKED IMPEDANCE.

motor a DEVICE that converts electrical energy into motion. Most motors convert to rotation, but linear motors convert AC energy directly into linear motion. See also UNIVERSAL MOTOR, INDUCTION MOTOR, SYNCHRONOUS MOTOR.

motorboating a form of PARASITIC OSCILLATION at low frequencies. Motorboating in audio amplifiers produces a noise like that of a motorboat engine. The cause is usually breakdown of a decoupling capacitor, and sometimes it is acoustic feedback.

moving-coil meter the basic DIRECT CURRENT measuring METER. A COIL attached to an indicating needle is suspended from springs or from a taut fiber in the field of a permanent magnet. The shape of the field is radial over a large part of its area because of the presence of a cylindrical magnetic core. Current flowing through the coil will cause torque that rotates the coil against the countering torque of the springs or suspension. The angle of rotation is proportional to the current through the coil over a wide range of angle. The current range can be extended by using shunts, voltage ranges can be read using series resistors, and ALTERNATING CURRENT can be measured by adding a RECTIFIER bridge.

moving-coil microphone a MICROPHONE in which the sound waves affect a moving COIL. The sound waves strike a diaphragm to which the coil is attached. The vibration of the coil then generates a signal voltage, because the coil is placed in the field of a permanent magnet.

moving-coil pickup a type of PICKUP for the groove type of disk recording. The stylus of the pickup is attached to a small COIL located in the field of a permanent magnet. The vibration of the stylus causes vibration of the coil, inducing a signal. Because both coil and magnet must be small, the signal amplitude is low, being measured in microvolts, and the output of such a microphone requires considerable amplification. The OUTPUT IMPEDANCE of the microphone is also low, however, making it easier to avoid picking up interference.

moving-iron meter a METER that makes use of the movement of a SOFT MAGNETIC MATERIAL. The magnetic material, called the ARMATURE usually forms part of a FLUX (2) path in a soft MAGNETIC core, and is suspended from springs and attached to an indicating needle. A COIL is wound around the magnetic core, and CURRENT in this coil will cause flux. The flux will cause a force to act on the armature, pulling it into line with the flux. This force is opposed by the spring suspension, so the angle of the turning of the armature is related to the amount of current.

The instrument is nonlinear unless the armature is suitably shaped. The advantage of a moving-iron instrument is that it measures true RMS values of signals whose frequencies are within its range. The instrument cannot be used for high-frequency signals because of the high inductance of the coil. See also HOT-WIRE INSTRUMENT, MOVING-COIL METER.

moving-iron microphone a MICROPHONE that uses the moving-iron principle. The vibrations of a diaphragm affect a soft magnetic armature that is placed in a FLUX (2) gap of a permanent magnet. A COIL wound around the magnet body converts variations of flux into signals.

moving-iron pickup a popular form of phonograph pickup that uses the moving-iron principle. The vibration of a stylus moves an ARMATURE that is part of a FLUX (2) circuit for a permanent magnet. A COIL wound around the magnet has a signal induced by the flux variations. The output is a few millivolts, and the impedance is low enough to avoid interference problems. The linearity depends greatly on the design of the magnetic circuit.

MSI *abbreviation for* medium scale integration, the scale of integration in which the number of active components on one chip is in the range of 10 to 1,000. See LSI, VLSI.

MTBF see MEAN TIME BETWEEN FAILURES.

MTI radar moving-target indicator radar, a system that uses cancellation of returns from fixed targets so that only moving objects are displayed. Modern MTI systems use computer techniques to carry out the cancellation and display. See also CANCELLATION CIRCUIT.

multiplexer 1. a CIRCUIT that permits simultaneous TRANSMISSION of several signals over a single channel communication system (a *multiplex* system). This can denote a CARRIER that is modulated in two ways, but more commonly denotes a circuit using PULSE methods. The original signals can be recovered by a DEMULTIPLEXER circuit. A single-channel system is sometimes called a *simplex* system. 2. (in instrumentation systems) a hybrid digital/analog device in which one of a number of analog signals is selected by means of a binary address and routed to a single output. 3. (in logic circuits) a device in which one of a number of digital inputs is selected by means of a binary address and connected to a single output.

multiplier a CIRCUIT or DEVICE that carries out a multiplying action. An ELECTRON MULTIPLIER multiplies the number of electrons in a beam. A FREQUENCY MULTIPLIER produces a harmonic of its input signal. An analog multiplier produces an output whose amplitude is proportional to the product of the amplitudes of the inputs. A *digital multiplier* carries out the arithmetic operation of multiplication on a pair of binary numbers.

multistable (of an electronic COMPONENT or CIRCUIT) possessing more than one stable state, like the old gas-filled counter tubes and more modern RING-COUNTER circuits.

multistage (of an electronic CIRCUIT) possessing more than one unit, particularly of amplification.

multivibrator a form of oscillator CIRCUIT for nonsine waves. The oscillator uses two INVERTING AMPLIFIERS that are crosscoupled. This means that the output of each amplifier is connected to the input of the other. If both connections are made through capacitors, the multivibrator is free-running and will generate a square wave for as long as power is present. If one connection is direct, and the other through a capacitor, then the circuit is MONOSTA-BLE. With two direct couplings, the circuit becomes a FLIP-FLOP. See Fig. 53.

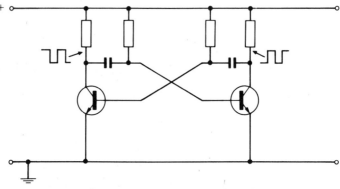

FIG. 53. **Multivibrator.** The circuit is shown here in its astable form. This circuit is a fundamental type of SQUARE-WAVE generator.

Mumetal a nickel-iron-copper alloy that has high PERMEABILITY and low hysteresis losses. Mumetal is used for MAGNETIC SCREEN-ING.

mush the NOISE in a radio receiver that is being operated with too weak a CARRIER signal.

muting the suppression of a SIGNAL by any DEVICE or CIRCUIT that shuts off unwanted signals. One application is automatic muting of interstation noise in FM receivers. The muting circuit comes into operation when the carrier strength falls below a preset limit, and this is used to suppress the audio output, thus eliminating the loud rushing noise that otherwise is heard before a signal is tuned in.

mutual conductance, see TRANSCONDUCTANCE.

mutual inductance the quantity that relates the MAGNETIC FLUX in one inductor to the current in another nearby inductor. Symbol: M; unit: henry. When mutual inductance exists between inductors, the inductors are said to be coupled, and the amount of COUPLING is determined by the value of mutual inductance compared with the individual SELF-INDUCTANCE of the inductors and by the presence of FERROMAGNETIC material. See also COUPLING COEFFICIENT.

Mylar *Trademark.* a type of polyester film, used as an INSULATOR particularly in CAPACITORS, and in photography and recording tapes.

N

NAND gate a type of LOGIC GATE. The gate output is logic 1 unless all its inputs are at logic 1, when its output is logic 0. The NAND gate is easy to manufacture and extensively used in small-scale integrated circuits. The advantage of using NAND gates is that any gate circuit can be built up from NAND gates alone. The same is true of NOR gates. See Fig. 54. See also AND GATE, OR GATE, NOT GATE, KARNAUGH MAP, INVERTER (1).

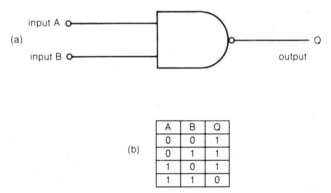

A	B	Q
0	0	1
0	1	1
1	0	1
1	1	0

FIG. 54. **Nand gate.** (a) NAND gate symbol. (b) Truth table.

nano- *combining form* denoting one billionth, as in nanosecond and nanofarad. Symbol: n.

narrow-beam antenna an antenna with a narrow RADIATION PATTERN and negligible SIDE LOBES. A narrow-beam antenna has a large GAIN in its aimed direction, compared with that of a simple DIPOLE. This can make for good reception even in noisy conditions.

natural frequency or **resonant frequency** the frequency of natural RESONANCE of a circuit or any other system that can oscillate. Any system that is set into oscillation by a brief stimulus will oscillate from then on at its natural frequency. An OSCILLATOR is most stable when the stimulus is so small that the frequency of oscillation is the natural frequency for the network.

A capacitor presents a high impedance to low-frequency alternating current, but has a small impedance at high frequency. The opposite is true for an inductor, which has a large impedance at high frequencies owing to the opposing voltage generated by rapid changes of current.

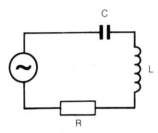

FIG. 55. **Natural frequency.** See this entry.

In a circuit containing inductive, capacitive, and resistive devices, as shown in Fig. 55, the impedance seen by the voltage generator will be large at high frequencies because of the inductor, and also at low frequencies because of the capacitor. At some intermediate frequency there will be a minimum impedance, and this is called the *resonant frequency* or *natural frequency*. In the circuit in Fig. 55, which is called a *series resonant circuit* resonance will occur when the angular frequency of the voltage generator ω is:

$$\omega = \frac{1}{\sqrt{LC}}$$

Resonance occurs at this frequency when the inductive and capacitive impedances are equal and opposite, so that they exactly cancel. Under these conditions the impedance is real and has the value R as though the circuit contained only a resistor.

A circuit excited into oscillation by an impulse will resonate until all the energy contained within the circuit has been lost through damping, in this case resistance.

Resonant circuits are often used to select signals near their resonant frequency. They are also used to construct oscillators.

See also RINGING.

NBS *abbreviation for* National Bureau of Standards.

n-channel a path of current constructed from an N-TYPE doped semiconductor. The term is usually applied to FIELD-EFFECT TRANSISTOR construction, in which the width of the channel is controlled by altering the BIAS on the surrounding material by means of a control electrode known as the GATE. See also P-CHAN-NEL.

negative (of a material) possessing an excess of ELECTRONS. The term is applied to any material in which there is a surplus of electrons over protons, whether these electrons are free or bound. Note that an n-type semiconductor is not negative in this sense, because the number of electrons is exactly balanced by the number of HOLES. The distinguishing feature of an n-type semiconductor is that many of the electrons are free to move and can be used as current carriers. Compare POSITIVE.

negative bias the application of a steady negative voltage or potential to a circuit or active device, for example, the GATE of a FET the base of a transistor, or the GRID of a vacuum tube or CATHODE-RAY TUBE to which a negative potential is applied in order to control the flow of electrons. A sufficiently large negative bias will prevent electron flow, and this amount of bias is called CUTOFF. Compare POSITIVE BIAS.

negative feedback the process by which a portion of the output of an amplifier is fed back to the input in such a way as to diminish the total input signal. This reduces the effective gain of the amplifier, but markedly increases its input impedance and reduces its output impedance. In addition, linearity is greatly improved. A system is said to be linear when its output is related to the input by an expression of the form:

$$S_{\text{out}} = CS_{\text{in}} + \mathbf{B}$$

when C and B are constants.

Negative feedback takes several forms. The feedback may be

proportional to either the output voltage or the output current, and it may be presented to the input either in series or in parallel with the signal to be amplified. There are, therefore, four basic types of negative feedback: *series voltage feedback, parallel voltage feedback, series current feedback* and *parallel current feedback.*

A further advantage of using negative feedback is that for an amplifier with high OPEN-LOOP GAIN, such as most OPERATIONAL AMPLIFIERS, the use of negative feedback means that the circuit gain is determined solely by the external, or feedback, component values. See Fig. 56. Compare POSITIVE FEEDBACK.

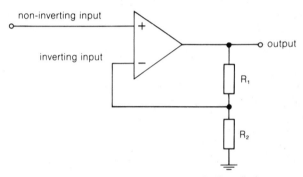

non-inverting input

inverting input

output

R_1

R_2

FIG. 56. **Negative feedback.** The circuit method applied to a general-purpose amplifier. The circuit assumes the use of a balanced amplifier. If gain without feedback is very high, then gain with feedback is approximately $(R_1 + R_2)/R_2$. This is the inverse of the division ratio of the resistor chain.

negative ion an atom that has gained one and sometimes more ELECTRONS and is therefore negatively charged. Because of the charge, the ion will experience force in an ELECTRIC FIELD and if free to move will do so. Compare POSITIVE ION. An ion should not be confused with a HOLE.

negative logic a logic voltage system in which the more negative voltage represents LOGIC ONE. Negative logic is seldom used today. Compare POSITIVE LOGIC.

negative resistance an unstable condition in which an increase of voltage across two points is accompanied by a decrease of current between the points. The term 'negative' is used because this resistance is the opposite of the normal ohmic resistance. Any device that can produce negative resistance over a range of voltage levels can be used as an OSCILLATOR. If negative resistance can occur in an AMPLIFIER under certain circumstances, the amplifier will be unstable under those conditions.

negative temperature coefficient (NTC) a figure that expresses the amount by which a quantity decreases as ambient temperature increases. NTC capacitors are used to stabilize RF oscillator frequency, and NTC resistors are used for similar purposes in RC oscillators. Compare POSITIVE TEMPERATURE COEFFICIENT.

neon a gaseous element used in display devices. The neon signal lamp consists of two small parallel ELECTRODES surrounded by low-pressure neon. The lamp glows orange-red at a POTENTIAL DIFFERENCE of about 70 V only after the gas has been ionized (see IONIZATION) by a higher potential difference, called the STRIKING POTENTIAL. When the gas is ionized, an increase of current through the gas causes a drop in the potential across the electrodes, so the characteristic is that of a NEGATIVE RESISTANCE. See Fig. 57.

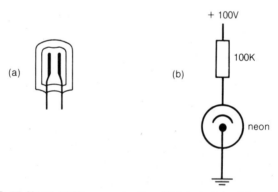

FIG. 57. **Neon.** (a) Neon construction. (b) A typical circuit for a neon indicator.

net loss the overall loss of SIGNAL usually expressed in DECIBELS in any device or network, whether active or passive.

network a circuit that consists of PASSIVE COMPONENTS such as resistors, capacitors, and inductors. Many types of electronic circuits can be divided into ACTIVE COMPONENTS such as operational amplifiers or transistors, and passive networks.

neutral 1. being grounded, or connected neither to positive nor to negative potential. 2. *n.* the return line of an AC power supply.

neutralization a type of connection or set of connections made in an amplifier STAGE in order to prevent OSCILLATION. A neutralizing connection generally consists of a capacitor connected between two stages of amplification and is intended to counteract the effects of STRAY CAPACITANCE by shifting the signal phase in the opposite direction.

Neutralization is particularly needed when a TRANSISTOR is being used for tuned radio-frequency amplification at a frequency near the limit of its frequency response. See also UNILATERAL NETWORK.

neutron a neutral elementary particle found in the nucleus of an atom. The nucleus also contains positively charged particles called PROTONS. A neutron has no charge, and its rest mass is 1.67482×10^{-27} kilogram, 1839 times that of an electron.

newton the SI and MKS unit of force. Symbol: N. A newton is defined as the force required to impart an acceleration of 1 meter per second per second to a body of mass 1kg. In Earth's gravitational field, the downward force on a 1kg mass is approximately 9.81 N.

nicad cell a nickel-cadmium cell or battery, a rechargeable form of alkaline electrolyte cell, used as a power source in portable equipment.

Nichrome *Trademark.* an alloy of nickel, chromium, and iron that has fairly high RESISTIVITY and can be used at fairly high temperatures. It is extensively employed in heating elements and can be used for WIRE-WOUND RESISTORS that will be operated with high DISSIPATION.

node 1. a joining point in a NETWORK particularly a FILTER network, at which components are connected. See Fig. 58a. 2. a point in a STANDING WAVE pattern at which there is no wave motion.

NOISE

See Fig. 58b. The position of maximum wave motion is called the ANTINODE. Nodes and antinodes are particularly important in ANTENNA design, and in TRANSMISSION LINES.

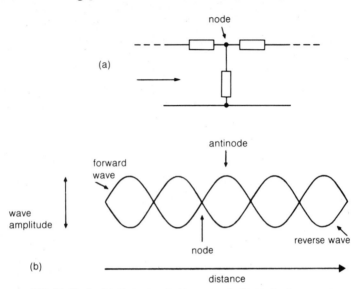

FIG. 58. **Node.** (a) Nodes in a ladder network. (b) Nodes in a standing-wave pattern.

noise any unwanted SIGNAL often signals of random frequency, phase, and amplitude. See also MAN-MADE NOISE, RADIO NOISE, SOLAR NOISE, IMPULSE NOISE, THERMAL NOISE, SCHOTTKY NOISE, WHITE NOISE.

noise factor the ratio of SIGNAL power to NOISE power. In TELEMETRY this factor is measured for a radio receiver that is tuned to a desired signal, first with no MODULATION so the noise output of the receiver can be measured, then with full modulation so the power of signal plus noise can be measured.

noise figure the NOISE FACTOR expressed in DECIBELS.

no-load condition the operation of an amplifier, oscillator, or other device without the LOAD normally be attached to it, that is, with no power taken from the circuit. An amplifier should be stable and not suffer damage under such conditions. An oscillator should show no frequency DRIFT.

nomogram a type of chart that enables the values of a variable to be estimated without resort to a mathematical formula. One form taken by a nomogram is a set of columns of values. By joining two columns with a straight line, values can be read from other columns that are intersected by the straight line. This usually can be as precise as is needed and it affords rapid estimates of quantities that would normally take much time to calculate. See Fig. 59. See also FORM FACTOR(2).

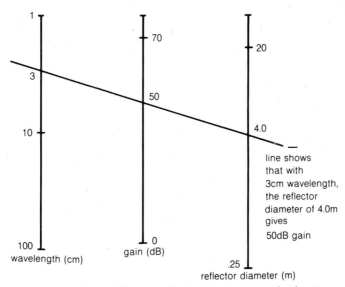

FIG. 59. **Nomogram.** The example shows a nomogram for the apparent power gain of a microwave parabolic reflector. The result is obtained merely by drawing a straight line between two condition values, 3cm wavelength and reflector diameter of 4 cm. The answer, 50dB gain, is then read off at the intersection. (In order to make the construction clear, only a few values have been shown.)

nondestructive test any form of testing that does not damage the test specimen. X-RAY examinations and ULTRASONIC scans are nondestructive tests.

noninductive (of a wire winding) being wound so that its INDUCTANCE is a minimum. The term is often applied to WIRE-WOUND RESISTORS, in which the required length of resistance wire is wound half in one direction and half in an opposite direction. The effect of this is to cancel most of the SELF-INDUCTANCE of the winding. For most purposes, the use of a metal FILM RESISTOR is a more satisfactory solution if inductance must be very low, but where high DISSIPATION or great accuracy is required, there may be no option but to use wire-wound construction.

nonlinear possessing a CHARACTERISTIC shape other than a straight line. The term is usually applied to an AMPLIFIER or other analog signal device in which a plot of signal out against signal in is not a straight line. Passive networks that do not contain metal-cored inductors are usually linear. Nonlinearity in an amplifier is a cause of HARMONIC DISTORTION and INTERMODULATION. See also NEGATIVE FEEDBACK. Compare LINEAR.

nonlinear distortion a type of distortion, such as HARMONIC DISTORTION and INTERMODULATION, caused by NONLINEAR behavior. The nonlinearity may be in an active device such as a transistor or integrated circuit, or in a passive device with HYSTERESIS such as an iron-cored inductor.

nonreactive (of a component) having only RESISTANCE. The term is applied to a RESISTOR or a TRANSMISSION LINE in which the PHASE of voltage is always the same as the phase of current under normal working conditions. See also IMPEDANCE, REACTANCE, RESISTANCE.

nonresonant (of a component or circuit) having a frequency characteristic with no sudden changes. A nonresonant component or circuit (see APERIODIC CIRCUIT) can be used at any frequency without risk of abrupt changes of phase or gain if the frequency is changed. See NATURAL FREQUENCY, RESONANCE.

nonsinusoidal (of a WAVEFORM) not being a sine wave but, usually, a pulse or square wave.

nonvolatile (of memory storage) not depending on a power supply for the storage of data, that is, any form of memory using MAG-NETIC or ELECTROSTATIC storage that will be retained when power is no longer applied. Compare VOLATILE. See also EAROM, EPROM.

NOR gate a form of logic gate whose output is at LOGIC ONE only when all its inputs are at logic 0. For all other combinations of inputs, the output is logic 0. The gate can be imagined as a combination of OR GATE and INVERTER (1), but it is always manufactured as a gate in its own right. Like NAND GATES NOR gates can be used in combination to provide the actions of any other types of gate.

normalized figure a figure that has been adjusted to fit a standard pattern. For example, RESONANCE curves are normalized around unity gain and resonant frequency. This means that the response is plotted with the peak gain shown as 1, and with the frequencies shown relative to the resonance frequency, rather than using actual measured values of gain and frequency for a given circuit. Normalization is encountered to a considerable extent in FILTER design and in PROPAGATION calculations. See Fig. 60.

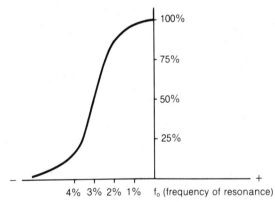

FIG. 60. **Normalized figure.** A normalized response curve, in terms of percentage gain and center frequency.

Norton's theorem the current equivalent of THEVENIN'S THEOREM for finding an EQUIVALENT CIRCUIT. It states that any two-terminal LINEAR NETWORK to which a signal is applied can be simulated by the behavior of a current GENERATOR in parallel with a resistor.

NOT gate see INVERTER (1).

n-p-n transistor a BIPOLAR TRANSISTOR with a BASE of p-type material, and an n-type EMITTER (1) and COLLECTOR. The n-p-n structure is the predominant one for silicon transistors.

NTC see NEGATIVE TEMPERATURE COEFFICIENT.

NTSC *abbreviation for* National Television Standards Committee. This is the committee that decides on technical standards for TV broadcasting in the US, notable for the 1951 color TV standard that is named NTSC after the committee. See also PAL, SECAM.

n-type (of a SEMICONDUCTOR material) being doped (see DOPING) with an electron-rich impurity. This causes the material to conduct predominantly by the movement of negative electrons. See also P-TYPE, TRANSISTOR, SEMICONDUCTOR.

nucleus the central part of an atom. The nucleus is positively charged and contains most of the mass of an atom.

null detector a DEVICE that detects when a signal has zero amplitude. It is often used in conjunction with the NULL METHOD of measurement to indicate balance. See also BRIDGE.

null method a measuring method in which adjustments are made in order to produce a NULL-DETECTOR reading. BRIDGES are the most widely known known null methods, because when a bridge circuit reaches its balanced condition, the signal or DC voltage across the detector is zero. This makes it unnecessary to use a detector that is calibrated in any way, and only high sensitivity is required. The size of the quantity being measured is calculated from the values of the other components in the bridge circuit and so is unaffected by any calibration error of the meter instrument.

numerical control the control of any SYSTEM by digital signals, often denoting the control of machinery by using digital signals from a computer.

Nyquist criterion a method of determining whether an amplifier will be unconditionally stable. This depends on whether the plot of a NYQUIST DIAGRAM encloses the point $-1, 0$ on the conventional normalized form of the diagram.

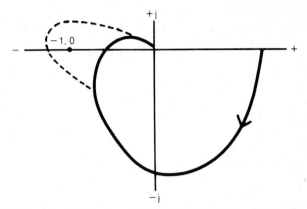

FIG. 61. **Nyquist diagram.** A typical Nyquist plot for an amplifier. This shows gain as a complex variable (see COMPLEX NUMBER) plotted on real and imaginary axes. The solid line, unconditionally stable, and the dotted line, unstable, represent the transfer function G, taking phase into account. If the plot indicates instability, the gain of the amplifier must be reduced around the critical frequencies.

Nyquist diagram or **vector response diagram** a plot of a NOR-MALIZED FIGURE of gain against phase shift for an amplifier. This plot shows the relationship between gain and feedback in an amplifier and is used as a measure of stability. See Fig. 61. See also NYQUIST CRITERION.

O

OCR see OPTICAL CHARACTER READER.

octave a doubling or halving of FREQUENCY. A change in response, such as response of a FILTER, is often given as the number of decibels of attenuation or amplification per octave.

off-line (of part of a computer system) disconnected from the computer system. An off-line device is normally active in the sense that power is applied to it, although it is prevented from communicating with the rest of the computer system. Compare ON-LINE.

offset a deviation from correct or normal VOLTAGE or CURRENT. The offset of an OPERATIONAL AMPLIFIER is the BIAS needed on one input to make the output reach ground voltage when the other input is at ground voltage.

ohm the SI unit of electrical RESISTANCE. The resistance in ohms between two points is defined as the DC voltage divided by the DC current flowing between the points. Symbol: Ω. See also CONDUCTANCE, REACTANCE, IMPEDANCE.

ohmic (of a material or a component) obeying OHM'S LAW having a constant value of RESISTANCE at a steady temperature. If the value of resistance varies with applied voltage or current, then the resistance is *nonohmic*. See also THERMISTOR.

ohmic contact an electrical contact with a steady value of CONTACT RESISTANCE.

ohmic loss a power loss caused by DISSIPATION in a resistor as distinct, for example, from radiation loss or HYSTERESIS loss.

ohmmeter an instrument for measuring RESISTANCE. Simple ohmmeters use a CELL of constant voltage to pass a CURRENT through the unknown resistor, and measure this current. The disadvantage is the nonlinear reading scale of a simple ohmmeter. More elaborate ohmmeters use measuring BRIDGE circuits. See also MEGGER.

ohm-meter the unit of RESISTIVITY. The resistivity of a material is the RESISTANCE between opposite forces of a cubic meter of the material at a steady temperature. Symbol: Ω m.

Ohm's law the principle that the RESISTANCE of a metallic conductor is constant at a constant temperature. For metallic and other ohmic conductors, we can use the relationship $V = R \times I$ between direct current and voltage because the quantity R (resistance) is a constant.

Nonohmic conductors do not have any constant resistance value, and the $V = RI$ relationship can be used only if the value of R under the conditions is known.

ohms-per-volt a figure used in comparing the sensitivity of DC VOLTMETER movements that shows the series resistance of the voltmeter for any voltage range. For example, if a meter is quoted as having 30 K per volt, then on its 10 V range, the resistance of the meter is 30×10 k-ohms.

oil-break switch a form of SWITCH whose contacts are immersed in oil. This suppresses ARCING when the contacts separate.

omnidirectional antenna an antenna whose RADIATION PATTERN is the same in all horizontal directions.

one-shot see MONOSTABLE.

on-line (of a part of a computer system) being switched on or connected to the remainder of the SYSTEM. Compare OFF-LINE.

opamp see OPERATIONAL AMPLIFIER.

open architecture see MOTHERBOARD.

open circuit a disconnection in an electrical CIRCUIT through which no CURRENT flows.

open-circuit impedance a measurement of the IMPEDANCE of a NETWORK. The open-circuit impedance at the input of the network is the impedance it has with the output-open circuit. The open-circuit impedance of the output is the impedance value it has when the input is open-circuit.

open-circuit voltage the output signal voltage of an OPEN CIRCUIT or device when no LOAD is connected.

OPEN-COLLECTOR DEVICE

open-collector device or **open-ended device** or **open-col-lector output** a digital device in which the active pull-up transistor of the output stage is omitted. Many digital device outputs are driven by a pair of transistors, and the output is held at 1 or 0 by a conducting transistor or MOSFET (see FIELD-EFFECT TRANSISTOR).

If a number of device outputs have to be connected together, this kind of active output is undesirable, since it may lead to an active low voltage being connected to an active high voltage, with disastrous consequences. To avoid this problem, open-collector devices are used. The output can only be made FALSE, which is a low voltage, or disconnected, and it is the responsibility of the user to provide a pull-up transistor so that in its disconnected state the output appears to be high. The pull-up resistor limits current and prevents damage when conflicting outputs are connected. Open-collector devices can degrade the speed and noise immunity of a logic signal, and this can be avoided by the use of a TRI-STATE DEVICE.

Any logical analysis can have one of two results: TRUE or FALSE. In digital electronics, HIGH and LOW voltages represent TRUE and FALSE states respectively. The exact voltage levels corresponding to TRUE and FALSE depend on the type of digital device used. However, TRUE is normally represented by $+5V$, and FALSE by $0V$. See also PUSH-PULL.

open-loop gain the GAIN of an amplifier when no FEEDBACK loop is connected.

open wire an uninsulated wire, as on an overhead line.

operating point a point on a graph that represents operating conditions. For example, on a graph of output voltage plotted against input voltage for a device (a TRANSFER CHARACTERISTIC), the operating point is the point on the graph that represents the static DC-bias condition with no signal applied. See Fig. 62.

operational amplifier (opamp) a very high-gain linear DC-cou-pled DIFFERENTIAL AMPLIFIER with a single-ended output. The gain of an opamp will typically be between 10^5 and 10^6. An opamp will have a high INPUT IMPEDANCE of 1 M.Ω or more, and a low OUTPUT IMPEDANCE of between 50 Ω and 5 kΩ. Most

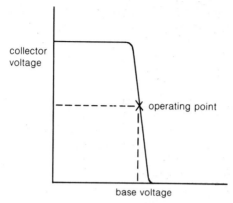

FIG. 62. **Operating point.** The operating point for a linear amplifier stage. The graph is a plot of collector voltage against base voltage for a transistor operating with a load. The operating point is chosen to be in the middle of the linear portion.

opamps use a split power supply, typically $\pm$ 15V. The symbol for an opamp is shown in Fig. 63(a), and it will be noticed that a pair of inputs are provided. The output becomes positive when the $+$ input is more positive than the $-$ input, and vice versa.

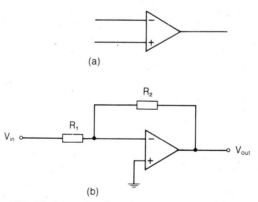

FIG. 63. **Operational amplifier.** See this entry.

OPPOSITION

An ideal opamp will have an output that can change instantaneously in response to the input. Practical opamps cannot do this; the rate at which the output can change is called the SLEW RATE.

Almost all opamp circuits can be analyzed using two simple rules:

I. The voltage between the + and − inputs is zero, to a very good approximation.

II. The inputs draw no current.

Consider the application of these rules to the circuit in Fig. 63(b):

(a) The + input is grounded, so Rule I states that the − input must also be at ground potential.

(b) Therefore, all of V_{in} appears across R_1, and all of V_{out} appears across R_2. Since Rule II states that the inputs cannot accept current, the current through R_1 must have an equal magnitude but an opposite sign to that through R₂.

In other words, $V_{in}/R_1 = -V_{out}/R_2$. The gain of this current V_{out}/V_{in} is therefore equal to $-R_2/R_1$.

This illustrates a common feature of opamp circuits. Because the intrinsic gain of the device and its input impedance are so high, Rules I and II can be applied, and the gain of an opamp circuit is determined solely by the input and feedback components.

opposition a SIGNAL that is a mirror image of another signal. Two signals of identical frequency and waveshape are said to be *in opposition* if their amplitudes at any instant are equal and opposite. The word *antiphase* is sometimes used, but for asymmetrical waves, a 180-degree phase shift does not produce an opposition, or inverted, wave.

optical axis the straight line connecting the centers of lenses in an optical SYSTEM.

optical character reader (OCR) a DEVICE that can scan printed characters and give a digital output to a computer. The scanning may be manual or automatic, and the output is usually in ASCII code form.

optical fiber a thin strand of transparent glass used to contain and transmit optical signals by total internal reflection. The small diameter of an optical fiber means that large numbers can be packed into the space occupied by a single conventional cable. Much higher data rates can be achieved using optical fibers in place of conventional conductors.

optical image an image in light, using dark and shade, or color of various degress of SATURATION. A TV CAMERA TUBE would convert an optical image like this into a CHARGE image, and so into an electron-beam image.

optimum bunching the bunching of electrons in a beam so that each bunch is separated by free space. See KLYSTRON, BUNCHER.

optoelectronics the study and the use of optical elements in electronic circuits. Optoelectronic devices such as PHOTODIODES and PHOTOTRANSISTORS can be used to build hybrid electronic-optical circuits, in which an electrical signal is connected to light pulses, passed along an OPTICAL FIBER and reconnected to electrical impulses at the other end. Some forms of signal processing can be carried out on a signal while it is in its optical form, for example, filtering, mixing, and switching. Recent developments of extremely fast optical switches have led to the possibility of optical computers being constructed, and research in this area is currently bearing fruit.

OR gate a LOGIC circuit that gives a LOGIC ONE output when any one, several, or all its inputs are at logic 1. The NOR GATE is the preferred type for manufacturing convenience.

The action of the OR gate is inclusive, meaning that the output will be logic 1 no matter whether there is one logic 1 input or more than one. For some applications it is necessary to exclude the case of more than one input being at logic 1, so the EXCLUSIVE-OR GATE is then needed. See Fig. 64.

orthicon an obsolete type of TV CAMERA TUBE first superseded by the IMAGE ORTHICON then by the VIDICON.

oscillating current a CURRENT that varies in a periodic manner with time.

oscillating voltage a VOLTAGE that varies in a periodic manner with time.

oscillation the production of a periodic VOLTAGE or CURRENT.

oscillator a CIRCUIT that produces an AC signal output from a DC supply. Any circuit that has a negative value of AC resistance over a range of VOLTAGES can be considered to be an oscillator. Con-

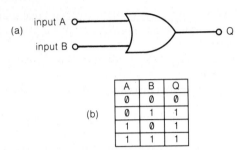

FIG. 64. **Or gate.** (a) Logic diagram. The logic of the gate is that the output is zero only when all inputs are zero. (b) Truth table.

necting a RESONANT CIRCUIT to this NEGATIVE RESISTANCE will result in a tuned oscillator whose output is close to a sine wave in shape.

If the oscillating conditions only just produce negative resistance over a short range, the waveshape will be a reasonably accurate sinusoid. If the negative resistance is large or if no tuned circuit is connected (see APERIODIC CIRCUIT), the output will be of square shape, as the oscillator output swings between BOTTOMING and CUTOFF. A few devices, such as TUNNEL DIODES have a negative resistance portion of their characteristic.

The other approach to oscillators is to produce a more controllable negative resistance characteristic by using an amplifier with POSITIVE FEEDBACK so that the output signal from the amplifier provides enough input signal to sustain oscillation. This approach is the basis of such common oscillator circuits as the COLPITTS OSCILLATOR, FRANKLIN OSCILLATOR, and HARTLEY OSCILLATOR.

oscilloscope see CATHODE-RAY OSCILLOSCOPE.

outgassing the removal of absorbed gas from the metal, ceramic, and glass of a VACUUM TUBE. This is done by use of an EDDY CURRENT HEATER while the tube is being evacuated. At later stages, outgassing can be assisted by running the tube filament and possibly passing anode current. Transmitting tubes use no GETTER to absorb traces of gas, so it is particularly important to perform the outgassing very thoroughly while the tubes are still connected to the vacuum pumps.

out-of-phase signal a SIGNAL whose phase is not in PHASE with respect to another signal. Out-of-phase is sometimes used to imply that the signal is INVERTED (see INVERSION), but the two shapes are not identical for signals other than sine waves.

output the final signal-delivering STAGE of an electronic circuit. Compare INPUT (1).

output gap the part of a MICROWAVE device from which the microwave signal is taken.

output impedance a COMPLEX NUMBER expressing the amplitude and phase relationship between voltage and current at the OUTPUT of a DEVICE or CIRCUIT.

output meter a METER that can be connected to the OUTPUT of a CIRCUIT to measure the power at the output.

output transformer a TRANSFORMER that is used at the OUTPUT of a CIRCUIT for COUPLING and IMPEDANCE MATCHING. Output transformers at one time were used in home audio amplifiers, but are seldom seen now because of developments in power transistors. The type of output transformer used in AMPLITUDE MODULATION is known as a *modulation transformer.*

overall efficiency the ratio of power supplied by a DEVICE to the power required to operate it.

overbunching the separation of bunched electrons after optimum bunching has been achieved. See KLYSTRON, BUNCHER.

overcoupling an excessive coupling (see CRITICAL COUPLING) between RESONANT CIRCUITS that will cause the amplitude-frequency graph to show a dip at the frequency of resonance. Compare UNDERCOUPLING.

overcurrent trip a safety DEVICE that opens a CIRCUIT when excessive CURRENT flows.

overdamped (of a RESONANT CIRCUIT or mechanical system) being DAMPED to an extent that prevents oscillations completely and causes a voltage or movement to die away too slowly. Compare UNDERDAMPED.

overdriven stage an amplifying STAGE with too large an INPUT signal. This causes distortion of the output and sometimes causes DEAD TIME so no output signal is produced.

overdubbing a separate recording of a piece of music from one instrument on a multitrack tape that can be used to replace a track made earlier.

over-horizon propagation the broadcasting of a radio signal over the visible horizon. PROPAGATION over the near horizon can be due to signal refraction in the lower atmosphere, but for greater distances the propagation is carried out by reflections from the IONOSPHERE. Modern satellites allow a different form of over-horizon propagation by the rebroadcasting of transmissions that are beamed up to a satellite from Earth.

overlapping gate a type of construction for CHARGE-COUPLED DEVICES.

overload take more than the rated power from a DEVICE or CIRCUIT. Overloading may cause a deterioration in the OUTPUT, such as an unacceptable waveform shape. Overloading may even cause damage to the device or circuit.

overload level the amount of output power at which OVERLOAD begins.

overmodulation the excessive AMPLITUDE MODULATION of a CARRIER so that the carrier amplitude is reduced to zero at each modulation trough. See Fig. 65.

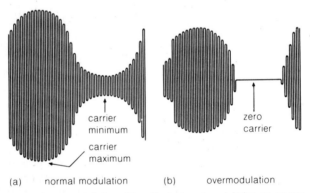

carrier
minimum

carrier
maximum

zero
carrier

(a)   normal modulation   (b)   overmodulation

FIG. 65. **Overmodulation.** The effect of overmodulation on an amplitude-modulated signal is to reduce the carrier amplitude to zero. This will cause severe distortion of the received signal.

overscan the application of excessive SCAN amplitude, particularly to a TV CAMERA TUBE. This is done to ensure that all of the active surface is used while the camera is working with a test signal. Using overscan avoids any danger of marking the normal picture area with a burned-in image.

overshoot an excessive transient portion of PULSE amplitude. At the leading edge of a pulse, the voltage may momentarily rise above the normal amplitude level of the top of the pulse. This transient is the overshoot. Compare UNDERSHOOT.

oxidation a chemical process in which an element is combined with oxygen to form a compound called an oxide. The oxides of metal are generally insulators, and the oxidation of silicon to produce nonconducting silicon oxide is an essential part of the production process of FIELD-EFFECT TRANSISTORS and INTEGRATED CIRCUITS.

The term is used widely in chemistry to denote any process in which electrons are removed from an ion, whether or not the process involves oxygen.

oxide masking the use of a layer of SILICON (IV) oxide to protect the underlying silicon. The oxide layer can be etched selectively with acid to permit the silicon to be treated.

P

PA see POWER AMPLIFIER, PUBLIC ADDRESS.

packing density 1. the number of ACTIVE COMPONENTS per unit area in an INTEGRATED CIRCUIT. 2. the number of molecules or atoms per unit volume in a crystalline material.

pad 1. a connection to an INTEGRATED CIRCUIT. 2. a computer network controller. 3. an ATTENUATOR circuit with a fixed division ratio.

padder see TRIMMER CAPACITOR.

pair a TRANSMISSION LINE with two conductors. The conductors may be parallel (twin-line), twisted, or in coaxial form.

paired cable a CABLE made from sets of twin conductors.

pairing a TV display fault that accentuates the appearance of scanning lines on the screen (see SCAN). Pairing occurs as the result of a synchronizing fault that prevents correct INTERLACE action. As a result, the even-numbered lines are in the same screen positions as the odd-numbered lines or very close to them. This makes the screen appear to have much lower vertical resolution. See Fig. 66.

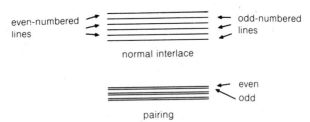

even-numbered lines   odd-numbered lines

normal interlace

even
odd

pairing

FIG. 66. **Pairing.** The effect of pairing on the lines that appear on a TV screen.

PAL *acronym for* phase alternate line, the standard European COLOR TELEVISION coding system. Like the SECAM system, PAL was devised to correct faults evident in the much earlier (1952) US NTSC system. Of these faults, the most troublesome was the effect of PHASE variations in the color SIGNALS which caused frequent and random color variations in the received picture.

In the PAL system, devised by Telefunken in Germany, the luminance (Y) signal is transmitted normally, and the color information is coded into two signals, the U and V signals. The U signal, with AMPLITUDE proportional to B-Y, is transmitted by modulating a SUBCARRIER in phase. The V signal, proportional to R-Y, is modulated on to the subcarrier at 90 degree phase. The V signal is inverted on alternate lines, giving the system its name. The phase of the synchronizing burst signal is also alternated to $+45$ degrees and -45 degrees relative to a reference OSCILLATOR.

At the TELEVISION RECEIVER the alternation of the burst phase is used to identify (by generating an IDENT SIGNAL) the phase of the V signal, and to invert it when necessary. Because of the inversion, phase changes in transmission can be canceled. Cancelation is accomplished by delaying the color signals for one line duration, and adding them to the color signals for the next line. The color signals are thus always averaged over two lines, in one of which the V signals will have been inverted during transmission.

The main disadvantage of both PAL and SECAM systems is that vertical RESOLUTION is halved compared with the NTSC system because of the averaging of the contents of adjacent lines.

PAM see PULSE-AMPLITUDE MODULATION.

panoramic receiver a RECEIVER that automatically tunes to a number of selected FREQUENCIES in sequence. Panoramic receivers are used to monitor international distress frequencies, or to monitor signals from a number of users of different frequencies.

paper capacitor a CAPACITOR whose DIELECTRIC is paper, usually impregnated with insulating oil. The metal conductors can be aluminum foil ribbons or aluminum metalized coatings (see MET-ALIZED PAPER CAPACITOR). The sandwich of conductors and paper is rolled in the form of a cylinder with another layer of paper. Values of capacitance in the region of 1 nF to 0.5 μF can be achieved in a reasonable size, and the voltage rating can be high, typically 100 V to 1 kV.

paper tape a now little-used method of permanent storage for computer data. The data are recorded by punching holes across the width of the tape, using up to eight preset positions. Each line of holes represents one character of a data file. The system is very slow and has been superseded by MAGNETIC DISK storage for all uses but that of maintaining the most important archives. See also MAGNETIC TAPE, MAGNETIC RECORDING.

parabolic reflector or **microwave dish** a concave mirror of parabolic cross section that is used as a receiving or transmitting antenna for radio waves in the form of a parallel beam. For transmission, the signal source, which is often the horn end of a waveguide, is located at the focus of the REFLECTOR. As a result, waves from the horn are reflected into a beam whose area of cross section is the same as that of the reflector. For reception, waves hitting the reflector are focused on to a small area at which the first receiver stage is placed. Small parabolic reflectors are now being sold for home use to pick up satellite broadcasts directly.

parallel (of a circuit or component) being connected in a way that makes it possible to share CURRENT across a common POTENTIAL DIFFERENCE. Resistors in parallel form a total resistance smaller than any of the individual values. Capacitors in parallel form a capacitance that is the sum of individual capacitance values. Cells in parallel with the same voltage output form a battery with higher current rating, but the same voltage output as a single cell.

parallel loading the loading (see LOAD) of a RESONANT CIRCUIT by connecting a resistor in PARALLEL. The lower this value of resistance, the greater the loading or damping (see DAMPED).

parallel-plate capacitor a CAPACITOR consisting of a pair of PARALLEL conducting plates separated by an INSULATOR called the DIELECTRIC. For such a structure, capacitance can be calculated from the dimensions of the plates and the thickness and relative PERMITTIVITY value of the DIELECTRIC. Most practical forms of capacitors are variations on this basic design.

parallel resonant circuit a RESONANT CIRCUIT that uses an inductor connected in parallel with a capacitor.

parallel-T network see TWIN-T NETWORK (1).

parallel transmission a connection between a COMPUTER and a PERIPHERAL that consists of a number of lines, one for each digital bit in the data unit. Parallel transmission is fast and is used for connecting the computer to disk systems. It is also used widely for connecting printers to computers, using the standard developed by Centronics Inc., and for digital network control, using the IEEE protocol developed by Hewlett-Packard.

parameter a quantity whose value is fixed only for one particular example. The parameters of a TRANSISTOR for example, include quantities such as the forward-current transfer ratio (current out/ current in). This parameter may be constant for a given transistor and set of conditions, and may have a different value under different test conditions, or in the next transistor that is tested.

parametric amplifier a form of AMPLIFIER used mainly for MICROWAVES. The parametric amplifier uses a circuit that contains a device, such as a VARACTOR for which a PARAMETER (capacitance in this example) is electrically variable. By using a signal to vary the parameter at a suitable frequency, a combination of amplification and frequency mixing can be achieved. The action of varying the parameter is called *pumping.*

paraphase amplifier an amplifier that converts SINGLE-ENDED signals, unbalanced signals whose sum amplitude is always zero, into balanced form (see BALANCED AMPLIFIER). The simplest form of paraphase amplifier is the transistor with equal loads in the emitter and the collector circuits. See Fig. 67.

parasitic antenna or **parasitic radiator** or **passive antenna** an ANTENNA element that does not receive or transmit signals and is used only to make an active antenna directional. See also REFLECTOR, DIRECTOR.

parasitic oscillations the unwanted OSCILLATIONS caused by POSITIVE FEEDBACK in an AMPLIFIER. The feedback is usually through STRAY CAPACITANCE or inductance and so is at a very high frequency. In some cases, such as audio amplifiers, the existence of parasitic oscillations may not be suspected as they can occur above the audio range. See also MOTORBOATING.

parasitic radiator see PARASITIC ANTENNA.

parasitic stopper a RESISTOR used to suppress PARASITIC OSCIL-

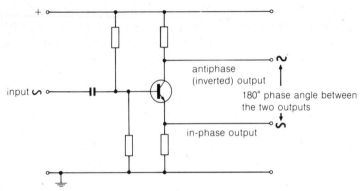

FIG. 67. **Paraphase amplifier.** A simple form of paraphase amplifier that produces both in-phase and inverted signals.

LATIONS. When parasitic oscillations cannot be stopped by using screening, changes in layout, or other methods of avoiding STRAY CAPACITANCE and INDUCTANCE effects, parasitic stoppers may be used. These are resistors connected into the path of the oscillation to cause damping (see DAMPED). Stoppers are usually resistors of 100 R to 1 K value wired as close to the electrodes of an active device as possible.

partition noise a form of NOISE in multielectrode devices. In the days when pentode vacuum tubes were used in receivers, partition noise arose because of random fluctuations in the electron currents to the various electrodes.

passband the range of frequencies a FILTER will pass from input to output without attenuation. The range usually is defined in terms of the frequencies that are attenuated by -3 dB as compared with the central frequency of the band. Compare REJECTION BAND.

passivation the protection of a SEMICONDUCTOR surface from unwanted contamination.

passive antenna see PARASITIC ANTENNA.

passive component a COMPONENT that cannot cause a POWER gain. Components such as resistors, capacitors, and linear inductors, including transformers, are all passive. Nonlinear action along with a power supply, as in a MAGNETIC AMPLIFIER can cause a component to be classified as an ACTIVE COMPONENT.

passive substrate an INSULATION layer on which a semiconductor can be formed or attached.

patch see HOOKUP.

patchboard or **plugboard** a board that contains sockets that are connected to circuits or devices. Experimental connections of CIRCUITS can then be arranged by connecting sockets with PATCH CORDS. See also BREADBOARD.

patch cord 1. a CABLE with a plug at each end, used for patching on a PATCHBOARD. **2.** an extension cable, especially one that allows a board taken from a circuit to be run externally for test purposes.

patching the making of temporary connections on a PATCHBOARD or BREADBOARD. This may be done to test a circuit idea by making measurements without having to construct a circuit or printed circuit board. Patching is feasible only for comparatively low-frequency circuits because of the large STRAY CAPACITANCES and INDUCTANCES of a patchboard system.

PC board see PRINTED CIRCUIT BOARD.

p-channel a conducting channel in a FIELD-EFFECT TRANSISTOR that uses P-TYPE doping. See also N-CHANNEL.

PCM see PULSE CODE MODULATION.

p.d. see POTENTIAL DIFFERENCE.

PDA see POST-DEFLECTION ACCELERATION.

PDM see PULSE-DURATION MODULATION.

peak clipping see CLIPPER.

peak factor the ratio of peak amplitude to ROOT MEAN SQUARE value.

peak forward voltage the maximum permitted voltage across a RECTIFIER when current flows in the forward direction. The rectifier may be a DIODE or a JUNCTION in a bipolar transistor.

peak inverse voltage the maximum permitted voltage across a RECTIFIER when bias is reversed, with no current flowing. This is an important quantity for a semiconductor junction such as a diode, because if the peak inverse voltage is exceeded, the diode or junction may break down. Unless the amount of reversed current can be limited, the diode or junction will be destroyed.

peak limiter see CLIPPER.

peak load a LOAD that causes maximum possible power to be dissipated from a circuit.

peak point the start of a NEGATIVE RESISTANCE region in the CHARACTERISTIC of a UNIJUNCTION or TUNNEL DIODE.

peak-riding clipper a type of self-adjusting CLIPPER. It charges a CAPACITOR to the value of the peak voltage of the signal and uses this voltage as a clipping voltage. In this way, brief interference pulses can be clipped, but fluctuations in the peak amplitude of the desired signal will not cause clipping of the signal.

peak-to-peak amplitude a measurement of AMPLITUDE particularly for an ASYMMETRICAL WAVEFORM that extends from the positive peak value to the negative peak value. See Fig. 68.

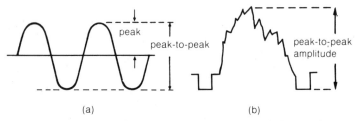

(a)   (b)

FIG. 68. **Peak-to-peak amplitude.** (a) For a symmetrical wave, the peak-to-peak amplitude is twice the peak amplitude. (b) For a pulse or any other asymmetrical waveform, the peak-to-peak amplitude is the only relevant measurement.

peak value an AMPLITUDE measured to one peak of a WAVEFORM from zero voltage or from a steady BIAS level.

peak white the level of AMPLITUDE of a TV signal that corresponds to the peak output of light on the screen.

pedestal a SQUARE WAVEFORM on which another pulse can be placed. The pedestal provides the starting voltage for the main pulse. See Fig. 69.

Peltier effect see THERMOELECTRIC EFFECT.

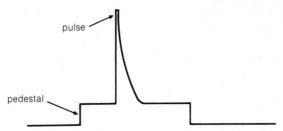

FIG. 69. **Pedestal.** A pulse superimposed on a pedestal. The addition of a pedestal is a method of ensuring, for example, that the pulse peak voltage cannot fall below the triggering level.

percentage modulation a measurement of linear AMPLITUDE MODULATION. Percentage modulation is the amplitude of the modulating signal divided by the amplitude of carrier signal and expressed as a percentage.

perfect crystal or **ideal crystal** a CRYSTAL in which each atom is precisely in position, with no particles of impurities to disturb the geometric and chemical perfection.

perfect transformer or **ideal transformer** a TRANSFORMER with no STRAY CAPACITANCE or stray inductance, no EDDY CURRENTS no HYSTERESIS in its core, and hence 100% COUPLING of signal and no LOSS.

period the duration of one CYCLE of a waveform, equal to the inverse of frequency.

periodic (of a WAVEFORM oscillation, coincidence, or effect) occurring at regular intervals.

peripheral any DEVICE external to the main system, that is, any device that can be connected to a computer, is controlled by the computer, but is not necessarily part of the computer. Examples include disk drives and printers.

Permalloy *Trademark.* a nickel-iron alloy, 30% to 90% nickel, that has high magnetic PERMEABILITY and unusually low HYSTERESIS loss.

permanent magnet a MAGNET made of material that has retained magnetism. The essential feature of a material for permanent magnets is high RETENTIVITY. The magnet can be demagnetized by heating or by applying a strong alternating field, which is then slowly reduced to zero. Permanent magnets are self-demagnetizing to some extent unless the flux path is kept closed by use of a KEEPER.

permeability the ratio of flux density (**B**) in a material to the magnetic field strength (**H**) that causes it. Symbol: μ; unit: henry per meter.

All materials have a value of permeability, but for most materials this value will be close to the value of the permeability of free space, about 1.25×10^{-6} henry/meter. For FERROMAGNETIC materials, the values of permeability can be much higher, but are variable. Since the value of permeability of a ferromagnetic material is proportional to the slope of the HYSTERESIS curve at any point, the value can vary from about the free-space value, at SATURATION, to a value many tens of thousands of times greater than the free-space value in the middle of the hysteresis curve. The permeability of a ferromagnetic material is usually given as the RELATIVE PERMEABILITY and the conditions under which it is measured must be stated.

permeability of free space the magnetic constant for a vacuum. The symbol is μ_0, and the value in modern units is approximately 1.25×10^{-6} henry/meter. The speed of light or any other electromagnetic wave in space is given by:

$$ c = \frac{1}{\sqrt{(\epsilon_0 \mu_0)}} $$

where c is the wave velocity and ϵ_0 is the PERMITTIVITY OF FREE SPACE. See also PERMITTIVITY, PERMEABILITY.

permeability tuning or **slug tuning** the tuning of a RESONANT CIRCUIT by varying the position of a core in a coil.

Permendur *Trademark.* a cobalt-iron magnetic alloy. The important characteristic of Permendur is its very high SATURATION flux density.

Perminvar *Trademark.* a nickel-cobalt-iron alloy. Its notable characteristic is a very constant value of RELATIVE PERMEABILITY over a wide range of applied magnetic field strengths.

permittivity a characteristic of an INSULATOR defined as the ratio of electric displacement (**D**) to electric field strength (**E**) in a material. Electric displacement is a quantity describing the behavior of a charge in a DIELECTRIC subjected to an electric field. Symbol: ϵ; unit: farad per meter. More practically, the permittivity value appears in the expression for the capacitance of a PARALLEL-PLATE CAPACITOR:

$$C = \frac{\epsilon A}{d}$$

in which C is capacitance in farads, A is plate area in square meters, d is plate spacing in meters, and ϵ is the permittivity of the dielectric. If the dielectric is air or a vacuum, then the value of permittivity is the PERMITTIVITY OF FREE SPACE ϵ_0.

permittivity of free space the value of PERMITTIVITY for a vacuum. Symbol: ϵ_0. The permittivity and PERMEABILITY of a medium determine the speed of electromagnetic waves in that medium. See PERMEABILITY OF FREE SPACE.

persistence the phenomenon of light being emitted from a PHOSPHOR screen for a time after the screen has been struck by an electron. See AFTERGLOW, LONG-PERSISTENCE SCREEN.

perveance a characteristic of a vacuum tube. Perveance measures the DC conductivity of the tube per unit area of cross section of the ANODE.

PFM see PULSE-FREQUENCY MODULATION.

phase a time difference between two identical WAVEFORMS of the same FREQUENCY. Phase difference is expressed either as a fraction of a CYCLE or, more usually, as a PHASE ANGLE. When the term is used to denote a single wave, it is the phase angle between current and voltage that is intended.

phase angle a measurement of the difference in PHASE of two waves. The phase angle between two waves, in degrees, is found by expressing the phase as a fraction of the cycle time and multiplying this figure by 360. If the phase angle is to be expressed in radians, then the time fraction must be multiplied by 2π.

phase constant the PHASE shift per unit length of line caused by a TRANSMISSION LINE.

phase control a method of using a THYRISTOR to control power in an AC circuit. In a phase-control system, the time at which the thyristor is switched on in each cycle can be varied with respect to the phase of the voltage.

phase-correction circuit a CIRCUIT that restores the original PHASE (of voltage relative to current) of a signal to compensate for the effect of another network.

phase delay the amount of PHASE shift per unit frequency of a wave.

phase deviation the difference in PHASE ANGLE of a phase-modulated wave compared with a standard, or reference, phase. See PHASE MODULATION.

phase difference the difference in PHASE between two waves expressed in terms other than PHASE ANGLE for example, time difference or fraction of a cycle.

phase discriminator or **phase-sensitive discriminator** a type of demodulator for phase-modulated signals (see PHASE MODULATION). The phase discriminator produces an output voltage that is proportional to the phase difference between the input signal and a reference signal.

phase distortion the unwanted PHASE DIFFERENCE between current and voltage of a waveform. Phase distortion is caused by the presence of reactive components in the current path.

phase inverter 1. in common usage, a signal waveform inverter, such as an INVERTING AMPLIFIER. 2. a circuit that changes the phase of a signal by 180 degrees.

phase-locked loop a CIRCUIT in which the PHASE of a LOCAL OSCILLATOR is locked to the phase of an incoming signal (see LOCKIN). This implies that the frequencies of the two signals are also equal. The locking action is carried out by using a PHASE DISCRIMINATOR to generate a signal proportional to the phase difference between the signals, and then using this signal to correct the oscillator frequency until phase equality is reached.

phase modulation (PM) the MODULATION of the PHASE of a CARRIER by a SIGNAL. The phase change of the carrier is made proportional to the amplitude of the modulating signal. The carrier amplitude remains constant, and the frequency of the variation is the same as the frequency of the modulating signal. The effect of phase modulation is indistinguishable from that of FREQUENCY MODULATION.

phase-sensitive detector see PHASE DISCRIMINATOR.

phase shift a change in the relative PHASES of two quantities. The term usually denotes a change in phase of voltage relative to current in a single wave.

phase-shift oscillator a form of OSCILLATOR that uses an INVERTING AMPLIFIER. The phase of the output signal is shifted by 180 degrees and connected to the input. For a sine wave, the phase shift has the same effect as an inversion, so the feedback is positive for the frequency at which the phase shift is 180 degrees. The output of the phase-shift oscillator is a sine wave only if the gain is maintained at the minimum level needed for oscillation. See Fig. 70.

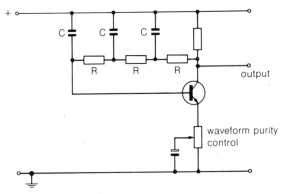

FIG. 70. **Phase-shift oscillator.** The waveform is sensitive to the GAIN of the STAGE which here is shown to be adjustable. For serious use, the gain would be controlled automatically.

phase splitter a CIRCUIT that produces a WAVE and its inverse, used mainly as a DRIVER for a BALANCED AMPLIFIER.

phase velocity the apparent speed of a WAVE peak in a group of waves in a dispersing medium. If the GROUP VELOCITY is less than the wave speed, the phase velocity is greater than the wave speed.

phasing 1. the adjusting of two signals so as to be in PHASE. **2.** the correct time adjustment of a synchronizing signal.

phonograph a sound reproducing system that uses a rotating plastic disk with a groove whose walls are mechanically engraved with the pattern of a sound wave. See also REPRODUCTION OF SOUND, COMPACT DISK.

phonon a mechanical wave QUANTUM used in connection with heat transfer and the propagation of acoustic waves.

phosphor or **fluorescent screen** a material used for coating CATHODE-RAY TUBE screens. Phosphors are metal compounds, usually sulfides or silicates of heavy metals, that have been activated by DOPING with other metals, notably silver. The effect is that incoming radiation, either as ultraviolet or electron beams, will cause light to be emitted by the phosphor. The emitted light is always at a lower energy level than that of the incoming radiation.

Phosphors can be manufactured for many colors of visible light and for infrared and ultraviolet by suitable choice of materials. The white phosphor for monochrome television use is a mixture of phosphors, with a predominantly blue cast. Phosphors for color TV use must be reasonably well matched, so the ratio of beam currents for equal brightness is not too great. The red phosphor is normally the least efficient.

photocathode a thin layer of material that emits ELECTRONS when struck by light in a vacuum. Photocathodes are based on materials such as antimony, cesium, rubidium, potassium, and sodium. These are easily prepared as thin films by evaporation, but can exist only in a vacuum. For use in PHOTOCELLS and PHOTO-MULTIPLIERS the photocathodes are made after the tube has been evacuated.

photocell or **electric eye** or **photoelectric cell** a DEVICE that will provide an electrical OUTPUT from a light INPUT (1). See also PHOTOCONDUCTIVE CELL, PHOTOELECTRIC EFFECT.

photoconductive cell a form of PHOTOCELL that uses a material whose conductivity is affected by light. Many semiconductors are photoconductive, but the materials used most frequently in photoconductive cells are cadmium sulfide and silicon, for small detectors, and antimony trisulfide or lead (IV) oxide, for TV camera tubes.

photoconductivity a variation of CONDUCTIVITY caused by light in which the conductivity of the material increases when illuminated.

photoconverter a form of ANALOG/DIGITAL CONVERTER that uses PHOTOCELLS one for each BIT of a binary number, and beams of light whose positions vary according to an input. The input can be from a rotating shaft or a sliding object, and the beams of light are obtained by shining a light through masks carried on the moving objects.

photocurrent an electric CURRENT produced in a PHOTOCELL by the action of light.

photodetector any electronic DEVICE that gives an electric OUTPUT when struck by light.

photodiode a light-sensitive SEMICONDUCTOR diode. The diode is operated with REVERSE BIAS and is contained in a transparent material. When illuminated by light, a reverse current flows. The current is small, to avoid damaging the device, but the speed of response of most photodiodes is rapid.

photoelectric effect the EMISSION of electrons from materials exposed to energetic radiation such as light. The velocity of the emitted electrons depends on light frequency, and their quantity on light intensity. For any photoemissive material, there will be a threshold frequency below which no emission can be obtained no matter how intense the radiation.

photoemission the EMISSION of electrons due to the impact of electromagnetic radiation. See PHOTOELECTRIC EFFECT.

photoemissive cell a form of photocell that makes use of PHOTOEMISSION.

photolithography a method of making PRINTED CIRCUITS and INTEGRATED CIRCUITS. A material is coated with PHOTORESIST, a light-sensitive liquid polymer. The material is then treated like a photographic plate and exposed to an image, often using ultraviolet rather than visible light. The effect of the exposure is to harden the photoresist selectively, so that washing the material will remove only the unexposed photoresist. This results in an image that is resistant to mild acids. The material that is not coated can then be etched with acid or otherwise chemically treated. Finally, the remaining photoresist can be removed with solvents or by heating.

The technique is stretched to its limits in the manufacture of integrated circuits, mainly because the distance between conducting lines in modern ICs is equal to a few wavelengths of light.

photomultiplier a combination of a PHOTOCATHODE and a set of secondary multipliers, or DYNODE CHAIN. The current output of a photomultiplier is much greater than that of a photocathode. However, the signal from a photomultiplier can be subject to NOISE as the DARK CURRENT is multiplied as is the desired photocurrent.

photon a discrete amount of radiated energy that behaves like a particle. The energy of a photon of a given radiation is found by multiplying the frequency of the radiation by Planck's constant, h.

photoresist the photosensitive liquid polymer used in PHOTOLITHOGRAPHY. The original photoresist was the gelatine-chromium (IV) oxide mixture that once formed the basis of the gum-bichromate photographic system.

photosensitivity the sensitivity of a material to light. In electronics, photosensitivity is usually applied to materials that give an electrical response to light.

phototransistor a TRANSISTOR provided with a transparent covering over the base region that renders it sensitive to light. The phototransistor is operated with no connection to the base. Base current is created by the PHOTOEMISSION effect of light striking the base region. The base current is then amplified by the normal transistor effect. The sensitivity is much higher than that of a photodiode, but the response time is longer.

phototube a tube, normally evacuated, that makes use of a PHOTOCATHODE to emit electrons. The electrons may be collected by an anode at a positive voltage, or multiplied by a DYNODE CHAIN as in a PHOTOMULTIPLIER.

photovoltaic cell a DEVICE that generates a VOLTAGE when exposed to light. The first photovoltaic cells used the element selenium, but these have been replaced by junctions formed between metals and a semiconductor, usually silicon.

photovoltaic effect the generation of a VOLTAGE at the junction of two materials when struck by light.

pi-attenuator an ATTENUATOR section with one series element and two parallel elements resembling the Greek letter π.

PIC *abbreviation for* proportional-integral control, a form of FEED-BACK control in which a signal proportional to the error amount is superimposed on a ramp obtained by integrating the output.

pickup a phonograph TRANSDUCER. It converts into an electrical output the mechanical vibration of a stylus that tracks the groove of a recording. The pickup is usually in the form of a plug-in cartridge that can be attached to a TONE ARM holdinf the cartridge and guiding it across the record surface.

The methods of converting vibration into electrical signals range from the PIEZOELECTRIC PICKUP for high output and relatively low quality, to the MOVING-COIL PICKUP for the lowest output and the highest quality. These systems may become obsolete with the introduction of digital COMPACT DISK systems.

pico- *combining form* meaning one trillionth. Symbol: p.

picture element or **pixel** the smallest part of a TELEVISION picture display that can be separately controlled. For TV reception purposes, this size is governed by the scanning frequencies and, for color, by the color system and aperture grid dimensions.

picture frequency the rate of transmission of a complete picture in television. See FRAME, RASTER.

picture noise the effect on a picture of a noisy signal. NOISE in a TV signal causes a speckled appearance on the picture. On a color signal, noise is most noticeable on areas of saturated red.

picture signal the video COMPOSITE SIGNAL that carries all the picture information.

picture tube a television CATHODE-RAY TUBE. The tube uses a large, usually rectangular tube face and is very short. The deflection is obtained by using coils around the neck of the tube, often shaped to obtain the very large deflection angle required. The tube is operated at a comparatively high accelerating voltage of about 14 kV for monochrome, 24 kV for color signals. The picture intensity variations are obtained by using the incoming video signals to control the electron beam current. A single electron gun is used for a monochrome tube, three guns for color tubes.

piezoelectric crystal a CRYSTAL such as quartz or barium titanate, that produces a voltage when stressed and changes dimensions when a voltage is applied. The crystals are cut so that the effect is most marked between two parallel faces, and these faces are metalized (see METALIZING), usually with silver. The crystals can then be used as electromechanical TRANSDUCERS.

piezoelectric cutter a sound-recording DISK cutter. The audio signals are applied to a PIEZOELECTRIC CRYSTAL that carries a cutting stylus. The cutter is forced to cut a spiral groove in a disk because it is attached to a screw-cutting lathe. As the audio signal causes vibration of the stylus, the cut groove is modulated. Two cutters set 90 degrees apart are needed for stereo recording.

piezoelectric loudspeaker a type of treble loudspeaker, or TWEETER. The piezoelectric tweeter uses a PIEZOELECTRIC CRYSTAL attached to a small diaphragm. When an audio signal is applied to the faces of the crystal, the diaphragm is vibrated.

piezoelectric pickup a PICKUP for conventional phonograph records that uses a PIEZOELECTRIC CRYSTAL. The early crystal pickups used materials such as Rochelle salt that were unstable and often failed in hot moist weather. Later types used ceramic cartridges, with materials such as barium titanate. These produced outputs in the tens of millivolts, but never with the linearity that can be obtained by using pickups of the magnetic type, such as MOVING-IRON PICKUPS or MOVING-COIL PICKUPS.

piezoelectric strain gauge a method of obtaining an electric signal proportional to strain in a material. The piezoelectric strain gauge uses a PIEZOELECTRIC CRYSTAL attached with epoxy resin to a part of a structure. Changes in the stress in the structure appear as voltage signals from the piezoelectric material.

pinch-off the reduction of CURRENT to zero in the channel of a FIELD-EFFECT TRANSISTOR. Because of the bias at the GATE (1), pinch-off occurs when the channel contains no free conductors.

pinch roller see CAPSTAN.

pincushion distortion the DISTORTION of a TV picture from a rectangular shape to a shape in which the sides curve inward. Compare BARREL DISTORTION.

p-i-n diode a DIODE with a thin layer of intrinsic, that is, high resistance, semiconductor between the P-TYPE and the N-TYPE layers. It is used as a microwave source (IMPATT diode) and as a photodiode (depletion layer diode).

pinout a chart showing the pin connections for an INTEGRATED CIRCUIT.

PIPO register a parallel-in, parallel-out REGISTER constructed from a connected chain of FLIP-FLOP units. Each unit has a separate input and a separate output. The units can load inputs and deliver outputs together, but there is no shifting of bits from one unit to another except by external connections. See Fig. 71. Compare PISO REGISTER, SIPO REGISTER, SISO REGISTER, FIRST-IN/FIRST-OUT BUFFER, FIRST-IN/LAST-OUT BUFFER.

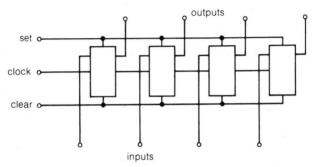

FIG. 71. **PIPO register.** The arrangement of a typical PIPO register, showing four bit units.

Pirani gauge a type of gauge for measuring low gas pressures in a partial vacuum. It operates by measuring the heat conduction through a gas and is suitable mainly for the lower levels of vacuum. See also IONIZATION GAUGE.

PISO register a parallel-in, serial-out REGISTER constructed from a connected chain of FLIP-FLOPS used for converting data from parallel to serial form. A PISO register can accept one bit of a parallel input at each unit, but outputs are obtained in serial form from one end only by clocking (see CLOCK) the register. See Fig. 72. Compare PIPO REGISTER.

pixel see PICTURE ELEMENT.

planar process a process for forming SEMICONDUCTOR junctions. It uses a layer of silicon oxide on a surface of pure silicon to ensure that diffusion processes affect only semiconductor surfaces that have been exposed by etching away the oxide layer.

237

PLANE POLARIZATION

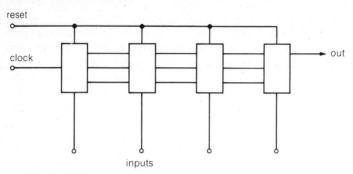

FIG. 72. **PISO register.** The arrangement of a typical PISO register showing four bit units.

plane polarization a type of polarization in which waves of light or other electromagnetic radiation have an electric component that is always in a fixed direction, the plane of polarization. See POLARIZED (1).

plan position indicator (PPI) a type of RADAR display in which the position of the receiver is indicated by the center of the display tube. Distances from the radar receiver are proportional to distances from the center of the tube face, and directions are read from a scale of compass positions around the tube face.

plasma see IONIZATION.

plasma display a form of ALPHANUMERIC display that uses the glowing gas of a plasma (see IONIZATION).

plasma oscillations the OSCILLATIONS of the current CARRIERS in a plasma (see IONIZATION). These cause severe radio-frequency interference.

plastic-film capacitor a capacitor of the rolled type that uses plastic film in place of paper (see PAPER CAPACITOR). The main types are polyester (see MYLAR) and polystyrene.

plate see ANODE.

plateau a portion of a graph that remains horizontal over a considerable range of values.

plated magnetic wire a wire of nonmagnetic material such as copper that is covered with a magnetic coating deposited by ELECTROPLATING.

238

plated-through connection a connection from one side to the other in a double-sided PRINTED CIRCUIT BOARD. A hole is drilled through the board, and metal is deposited on the sides of the hole to link the copper strips on each side of the board.

plug a male connector for connecting signals or power. Compare SOCKET.

plugboard see PATCHBOARD.

plug-in (of part of a circuit) being connected or disconnected by a PLUG and SOCKET.

plumbicon a form of VIDICON that uses a lead (IV) oxide photoconductor.

plumbing a slang term for MICROWAVE signal connections through waveguides.

PM see PHASE MODULATION.

p-n junction an area of contact between a P-TYPE and an N-TYPE semiconductor within a CRYSTAL. The JUNCTION cannot be created simply by bringing two materials together, because there must be no discontinuity in the crystal structure at the junction.

At such a junction, a rectifying effect will be obtained. Current will flow easily in the forward direction of the junction, with the p-type material at a higher positive voltage than the n-type material. No current flows until a threshold voltage is reached, after which the current-voltage graph is exponential. In the reverse direction, current is negligible until the voltage reaches a level at which the junction breaks down.

p-n-p transistor a TRANSISTOR formed of a P-TYPE material for the EMITTER (1) and COLLECTOR regions, and an n-type for the BASE. In use, both the collector and the base must have negative bias with respect to the emitter. See also N-P-N TRANSISTOR.

point-contact diode an early type of semiconductor DIODE construction. A point-contact diode used a metal wire ending in a sharp point, which made contact with a small amount of semiconductor material. The diode was used mainly for demodulation of high-frequency, low voltage signals.

poison an impurity in a material that prevents normal action. Compare DOPING which is a deliberate addition of an impurity in order to obtain desirable effects.

POLARITY

polarity 1. a variety of charge, either positive or negative. 2. a magnetic direction, either north or south. 3. a method of connecting certain components, for example, electrolytic capacitors, intended to avoid damage.

polarized (of an electromagnetic wave, electronic component, or connector) 1. having the electric field portion of an electromagnetic wave exhibiting a defined amplitude in a defined direction. This is taken as the direction of polarization. In a few cases (see TM WAVE), the direction of the magnetic portion of the wave is considered. 2. having a fixed charge, such as that possessed by an ELECTRET or requiring a potential of stated polarity on each of two connectors, like an ELECTROLYTIC CAPACITOR. 3. making connection between a plug and a socket in such a way that there is only one way to insert the plug in the socket.

pole 1. a point on the complex S-plane (see POLE-ZERO DIAGRAM). 2. a center of charge or magnetism. 3. a terminal, especially a cell or battery terminal.

pole-zero diagram a graph showing the voltage responses (*poles*) and current responses (*zeros*) of a circuit. For example, the natural response of a second-order system is determined by the roots of its characteristic equation. In general these roots are COMPLEX NUMBERS.

As an example, consider the locus of roots for a series RLC (resistor, inductor, capacitor) circuit as R varies. The characteristic equation of such a circuit is

$$LS^2 + RS + 1/C = 0$$

and the roots are

$$S = -R/2L \pm j(1/LC - R^2/4L^2)^{\frac{1}{2}} = a \pm j\ \omega$$

where S = a complex frequency, $j = \sqrt{-1}$, and ω is angular frequency. When $R = 0$, $S = \pm j\omega_n$, defining values of the undamped NATURAL FREQUENCY. When $R > 0$, a real part of S appears, and the values of S are complex conjugates.

The governing equation of the RLC circuit is

$$L \frac{dI}{dt} + RI + \frac{1}{C} \int I dt = V$$

If the current I is exponential, that is, $I = I_0 e^{St}$, this becomes

$$\mathbf{L}\mathbf{S}I + RI + \frac{1}{\mathbf{S}C}I = V$$

and we can define impedance $\mathbf{Z(S)}$ as

$$\mathbf{Z(S)} = \frac{V}{I} = \mathbf{S}L + R + \frac{1}{\mathbf{S}C}$$

The magnitude of $\mathbf{Z(S)}$ can be plotted as a height above the $\mathbf{S}$ plane. Typically, the impedance function plotted in three dimensions (α, ω, and $|\mathbf{Z(S)}|$) has the appearance of an exponential-sided cone. The tips of the cones, called poles, signify components of the natural voltage response of a circuit. Points at which the magnitude of $\mathbf{Z(S)}$ is zero, called zeros, signify components of the natural current response of a circuit. The pole-zero concept is a powerful tool for determining the natural forced behavior of any linear circuit.

polyester-film capacitor a rolled capacitor that uses polyester film in place of paper. See PAPER CAPACITOR.

polyphase supply an electrical supply that uses more than two PHASES. The usual type is three-phase, in which the currents in the three lines are 120 degrees out of phase with one another.

polyphase transformer a TRANSFORMER with several windings that can be connected to a polyphase supply.

polystyrene-film capacitor a rolled capacitor using polystyrene film in place of paper. See PAPER CAPACITOR.

port 1. an INPUT/OUTPUT circuit, particularly one for a computer system. A port must include storage, because there can be no certainty that the input or output will take place at a time when it can be used. In addition, a port includes a method of interrupting (see INTERRUPT) the action of the MICROPROCESSOR so as to trigger the reading of an input. **2.** a terminal on a device.

position sensor a device, sometimes based on the HALL EFFECT that can sense the position of a mechanical device. Other examples include optical-shift position sensors and ultrasonic proximity detectors.

positive (of a material) having a deficiency of ELECTRONS. The term is applied to any material in which the number of electrons is less than the number of protons. Compare NEGATIVE.

positive bias the application of a steady positive potential to a circuit or active component terminal such as the GATE (1) of a field-effect transistor. Compare NEGATIVE BIAS.

positive feedback or **regeneration** the process of sending a portion of the output back to the input of an amplifier in such a way as to reinforce the input signal (see FEEDBACK). Positive feedback increases gain, narrows bandwidths, increases distortion, and can cause oscillation. Compare NEGATIVE FEEDBACK.

positive ion an atom that has lost one or more ELECTRONS and is therefore positively charged. Because of the charge, the positive ion will experience a force in an electric field opposite in direction to that experienced by an electron. Compare NEGATIVE ION.

positive logic the logic system in normal use, in which a positive voltage represents level 1, and zero or negative voltage represents level 0. Compare NEGATIVE LOGIC.

positive temperature coefficient (PTC) a figure that expresses the amount by which a quantity increases in value as temperature increases. Compare NEGATIVE TEMPERATURE COEFFICIENT.

post-deflection acceleration (PDA) the acceleration of an electron beam in a CATHODE-RAY TUBE after deflection of the beam. The method applies to electrostatically deflected tubes as used for oscilloscopes and other instruments. For high deflection sensitivity, the electron velocity must be low; for high brightness, the velocity must be high. In a PDA tube, the accelerating voltage as far as the DEFLECTION PLATES is low. After the deflection plates, the beam is accelerated to a high velocity so as to produce a bright trace.

The separation of the low-velocity and high-velocity regions can be very complete, as in the MESH-PDA TUBE system. An alternative is the SPIRAL PDA tube, in which an anode at the deflection plates is connected to the final anode by a thin spiral of high-resistance graphite. This enables a graduated accelerating field to be created between the end of the deflection region and the screen.

pot see POTENTIOMETER.

potential see ELECTRIC POTENTIAL.

potential barrier the difference in VOLTAGE between two points that forms an obstacle to the movement of a charged particle. A negative electrode, for example, forms a potential barrier to an electron.

potential difference (p.d.) the difference in ELECTRIC POTENTIAL between two points. Symbol: U or V or ΔV; unit: volt. Potential difference is the amount of work done per unit charge when a charge is moved from one point to another along any path.

potential divider a series arrangement of RESISTORS. A voltage is applied over the whole set, and fractions of the voltage can be obtained wherever two resistors are connected. The division formula is expressed as:

$$V_{out} = V_{in} \frac{R_2}{R_1 + R_2}$$

Capacitors or inductors can be used to form potential dividers for alternating signals. See Fig. 73. See also COLPITTS OSCILLATOR, COMPENSATED ATTENUATOR.

potential gradient or **electric field strength** the POTENTIAL DIFFERENCE between two points divided by the straight-line distance between the points. Unit: volts per meter.

potentiometer (pot) 1. a variable resistor having a moving contact that can be placed anywhere between the terminals of a resistive coil or carbon track. 2. an instrument for measuring ELECTRIC POTENTIAL using a tapped voltage divider.

powdered-iron core see DUST CORE.

power the rate of production or dissipation of energy. Symbol: P; unit: watt. In DC circuits, power is calculated from the product of

POWER AMPLIFIER (PA)

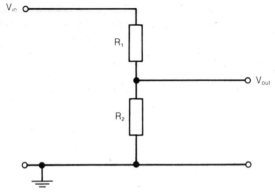

FIG. 73. **Potential divider.** See this entry.

POTENTIAL DIFFERENCE and CURRENT $V \times I$. For AC systems, the quantities must be measured in ROOT MEAN SQUARE terms and be in phase.

power amplifier (PA) 1. a STAGE that provides output POWER to a LOAD. **2.** the final RF AMPLIFIER that provides POWER to the antenna from a transmitter. **3.** the audio power amplifier stage that delivers the signal to the loudspeaker(s).

power component the in-phase component of current or voltage.

power efficiency the ratio of output power to input power, particularly for a TRANSDUCER.

power factor the fraction by which the product $V \times I$ for AC signals must be multiplied to obtain true POWER. The power factor is equal to the cosine of the phase angle between voltage and current.

power gain the ratio of POWER output to power input for any device or circuit.

power pack a circuit that converts AC power supply voltage into low-voltage DC for an electronic device. The conventional power pack will consist of an AC power transformer, RECTIFIERS and smoothing, possibly with voltage stabilization stages. See also RESERVOIR CAPACITOR.

power supply the source of POWER for an electronic circuit that can be AC power, battery, solar cell, or other methods.

244

power transistor a TRANSISTOR designed to handle large currents and dissipate heat efficiently. A power transistor usually has a large area of collector in good thermal contact with a metal casing. This allows the transistor to dissipate a large steady power, provided that the case can be cooled by conduction or convection. See HEAT SINK.

PPI see PLAN POSITION INDICATOR.

PPM see PULSE-POSITION MODULATION.

preamplifier (preamp) a STAGE or set of stages of low-noise voltage amplification, used in audio amplifiers in particular. In audio amplifiers, the preamp handles all the selection and mixing of inputs at various voltage levels. The system is arranged so that the output will be of about the same amplitude for any selected input. The output of a preamplifier is normally connected to the input of a POWER AMPLIFIER stage.

preemphasis a selective boosting of high audio frequencies. Preemphasis is used before transmitting FM signals, and before recording on tape or on disk. In all these systems the noise tends to be concentrated mainly at high frequencies. Using a preemphasized signal, the receiver or playback stages can cut treble or high frequencies (DE-EMPHASIS) and the accompanying noise. This restores the treble to its correct level and reduces noise.

preferred values a set of standardized figures between 1 and 10. These are based on a logarithmic scale and are used to select values for RESISTORS and CAPACITORS. The principle is that if a set of values is selected in a tolerance range of 20%, for example, then no value of resistance or capacitance can possibly be rejected. A nominal 2K2 resistor, for example, in the 20% tolerance range, could have a value of up to 2K64, or down to 1K76. The next value of 3K3 could be as low as 2K64, and the previous value of 1K5 could be as high as 1K8. A range of resistors or capacitors can therefore be manufactured, and values selected for 5% accuracy. The remaining components can then be selected to provide a set of 10% components, and the remainder can be considered as 20% tolerance. See Fig. 74.

pre-pulse trigger a TRIGGER PULSE for a CATHODE-RAY OSCILLO-SCOPE. In a radar transmitter, for example, a system in which a master pulse is generated and used to trigger all subsequent ef-

E 6 series	E 12 series	E 24 options
20%	10%	5%
1·0	1·0	1·0
		1·1
	1·2	1·2
		1·3
1·5	1·5	1·5
		1·6
	1·8	1·8
		2·0
2·2	2·2	2·2
		2·4
	2·7	2·7
		3·0
3·3	3·3	3·3
		3·6
	3·9	3·9
		4·3
4·7	4·7	4·7
		5·1
	5·6	5·6
		6·2
6·8	6·8	6·8
		7·5
	8·2	8·2
		9·1

FIG. 74. **Preferred values.** Three of the preferred number series for component values. Any multiple or submultiple of these numbers can be used.

fects, it is useful to provide a pulse that immediately precedes the main pulse. This makes it possible to trigger an oscilloscope timebase just before the main pulse arrives, so that the shape of the entire main pulse can be displayed. If the main pulse is used for triggering, its leading edge cannot be seen on the oscilloscope display.

Prestel *Trademark.* an information service for computers that makes use of the telephone network.

PRF see PULSE REPETITION FREQUENCY.

primary electron an electron obtained from a cathode of any type. Compare SECONDARY ELECTRON.

primary radiator the main element in an ANTENNA. For a TRANS-

MITTER the primary radiator is the electrically connected antenna element. For a receiving antenna, the primary radiator is the element connected to the RECEIVER.

primary service area the area of good reception around a TRANS-MITTER. The primary service area is served by the ground wave, and interference from reflected waves is small. This ensures a good standard of reception around the clock.

primary standard a national or international standard of measurement, such as the National Bureau of Standards current balance for the ampere. SECONDARY STANDARDS are calibrated from instruments of this type for everyday use, since primary standards may take a long time to set up and use.

primary voltage the AC or signal VOLTAGE across the primary winding of a transformer.

primary winding the WINDING to which the input signal for a transformer is connected. In the conventional iron-cored transformer, the primary winding is wound immediately next to the core. For modern toroidal transformers, primary and SECONDARY WINDINGS are often interchangeable. See also TERTIARY WINDING.

principle of superposition see SUPERPOSITION THEOREM.

printed circuit a CIRCUIT in which connections between components consist of copper strips on an insulated board. The components are introduced by drilling holes for terminating leads, threading the leads through, and soldering them to the copper strips. The pattern of connections on the board is printed on by PHOTOLITHOGRAPHY and the unwanted metal is etched away in ferric chloride or ammonium persulfate. Double-sided boards have a pattern printed on each side, with PLATED-THROUGH CONNECTIONS between each side. Connections to the board are made using EDGE CONNECTORS or by soldering sockets to the board. The advantages of printed circuits include uniformity, since the boards can be mass-produced, and ease of automated assembly.

printed circuit board (PC board) a board prepared by PHOTOLITHOGRAPHY ready for insertion of components. The board material is usually silicon-resin bonded paper or fiberglass, and the copper sheet is bonded to the board during manufacture of the board. The bonding must be sufficiently heat-resistant to allow DIP

247

SOLDERING of the entire board after the components have been threaded into place.

print-through a transfer of MAGNETISM from one layer of magnetic tape to another. Print-through causes the effect of *pre-* or *post-echo*. This means that a loud sound can cause a faint copy to be heard just before and just after the correct position on the tape. Print-through is minimized if tape is frequently wound from reel to reel.

probe 1. a lead for injecting or extracting signals, for example, the sharp-pointed or clip-ended lead of a test instrument such as a voltmeter, signal generator, or oscilloscope. 2. a conductor that is inserted permanently into a RESONANT CAVITY of a microwave device to inject or extract signals.

PROELECTRON code a British system of coding TRANSISTORS and other SEMICONDUCTOR devices. The PROELECTRON code uses letters and numbers. The first letter is a material code, with A = germanium, B = silicon. The second letter denotes the device type, such as C for audio voltage amplifier, F for RF amplifier. A last letter of X, Y, or Z is used to denote an industrial device. The number that follows is the maker's development number.

The advantage of the PROELECTRON code is that the type of device can be deduced from the code, unlike the US system of serial numbers.

program a set of instructions for a computer or a MICROPROCESSOR.

programmable capable of being controlled by a stored set of instructions, usually in number form.

PROM *acronym for* programmable read-only memory, a form of permanent NONVOLATILE memory for a computer. The data stored on the chip can be read by the computer, but no new data can be written while the chip is in the computer. The chip can be removed from the computer, erased, and reprogrammed if changes are needed. See also EAROM, EPROM, ROM.

propagation the spreading of a radio wave from its source to all the points at which it can be received.

propagation coefficient a quantity that measures the attenuation and phase change, per meter, caused by a TRANSMISSION LINE.

propagation delay the time delay between the PROPAGATION of a signal into a line and the reception of the signal at the other end. Also used of a broadcast signal to mean the (much shorter) delay caused by the time taken for the wave to move from transmitter to receiver.

propagation loss the energy loss from a signal beam in the course of PROPAGATION by radio, cable, or optical fiber.

propagation prediction a prediction of ionospheric weather conditions. Prediction of changes in the IONOSPHERE is possible with reference to the condition of the sun and past experience. In this way, predictions of propagation conditions for various wavebands can be made in advance, so important communications can be routed through channels that will be least affected by the expected changes.

proportional control a control system that makes use of NEGATIVE FEEDBACK. The output of the system is measured, and the result of the measurement is used to correct the controlling signal.

proportional counter a form of radiation detector. The detector is biased so that the amount of current produced is proportional to the energy of the ionizing radiation.

protective gap a spark gap used to protect high-voltage equipment. Any sudden voltage surge will cause a spark at the protective gap instead of causing breakdown of insulation or other damage to electronic circuits.

protective relay a RELAY that will disconnect a circuit if a fault occurs. See also CONTACT BREAKER, OVERCURRENT TRIP.

proton a stable, positively charged elementary particle found in the nucleus of an atom. The number of protons in an atom is equal to the number of electrons—equivalent to the atomic number of the element.

The nucleus of an atom also contains neutral particles called NEUTRONS. The charge of a proton is 1.6022×10^{-19} coulomb. This charge is of equal magnitude but opposite sign to that of an ELECTRON. The REST MASS mass of a proton is 1.67252×10^{-27} kilogram, 1836 times that of an electron.

proximity effect a form of INTERFERENCE between signal conductors. High-frequency signals are carried by currents that flow on the outsides of conductors. If two such conductors are close to one another, the repulsion of the electrons can cause the effective area of the conductors to be restricted, making the apparent resistance higher.

PTC see POSITIVE TEMPERATURE COEFFICIENT.

PTFE (*Trademark.* **Teflon**) *abbreviation for* polytetrafluoroethylene resin, a plastic material with very high resistivity, low dielectric loss, and low coefficient of friction. Almost unaffected by moisture, PTFE is used extensively in high-voltage insulation and in cable connectors.

PTM see PULSE-TIME MODULATION.

p-type (of a semiconductor material) being doped with ACCEPTOR material. P-type material conducts by HOLE mobility rather than by ELECTRON mobility. See also N-TYPE.

public address (PA) a sound-reinforcing system for large numbers of listeners that uses microphones, amplifiers, and loudspeakers. Problems of PA systems include echoes, acoustic feedback, lack of intelligibility, and NOISE.

pulling the disturbance of OSCILLATOR frequency by another signal at a nearby frequency. Unless an oscillator is very stable, it can be made to synchronize with a strong signal at a close frequency. Pulling is minimized by isolating an oscillator as much as possible from other circuits. This can be done, for example, by using a separate stabilized supply, encasing the oscillator section in metal, and connecting its output through a buffer stage.

pullup resistor a RESISTOR intended to BIAS an electrode to a positive potential, usually that of the supply positive terminal. Several types of logic integrated circuits need pullup resistors if their outputs are not connected directly to another input or if a number of outputs are connected.

pulsating oscillating in size, brightness, or other characteristic.

pulse a single cycle of abrupt change in VOLTAGE or CURRENT. The usual pulse form consists of a sharp rise of voltage, followed by a drop to the original level, the BASE LEVEL. In most systems, pulses are used in a pulse train, meaning that the pulses repeat at regular intervals. The intended shape of a pulse is usually rectangular, but due to effects of STRAY CAPACITANCES and inductances, this shape is modified to give the effects of OVERSHOOT, DROOP, and RINGING

(2). A perfect rectangular pulse would require no time to change its voltage value, but in practice the requirement to charge and discharge stray capacitance means that definite rise times and fall times can be measured. These times are usually measured between the 10% and 90% amplitude points. The width of the pulse is the time during which its amplitude exceeds the 90% level.

A pulse may be of current rather than of voltage. Pulse shapes other than rectangular are sometimes used, such as triangular and cosine-squared pulses.

pulse-amplitude modulation (PAM) a form of MODULATION based on PULSES at MICROWAVE frequency. The signal to be transmitted is used to modulate the amplitude of pulses in a pulse train. DEMODULATION is simple, employing a LOW-PASS FILTER is used, but this type of transmission is susceptible to interference.

pulse-code modulation (PCM) a form of MODULATION based on transmission of DIGITAL signals. Signal amplitude is sampled at intervals by an ANALOG/DIGITAL CONVERTER. The output of the converter is used to generate a series of pulses, so that a set of pulses is sent for each point on the waveform of the signal. At the receiver, these signals are decoded, and the signal is recovered from the digital-to-analog converter.

pulse discriminator a CIRCUIT that selects PULSES of a particular specification on the basis of their amplitude, width, or repetition frequency.

pulse-duration modulation (PDM) or **pulse-width modulation (PWM)** a form of MODULATION using a SIGNAL to modulate the width of PULSES. The width or duration of each pulse is proportional to the amplitude of the modulating signal at that instant. DEMODULATION can be carried out by a simple INTEGRATOR.

pulse-forming line a line or cable with a reflecting TERMINATION used to control the pulse width in a pulse generator terminated with a short circuit or an open circuit that will reflect a PULSE back along the line. The length of the line will determine the time needed for a pulse to be propagated down the line and then reflected back, and this time delay can be used to control the pulse width in a pulse generator circuit. An artificial LUMPED PARAMETER line is often used.

pulse-frequency modulation (PFM) a form of MODULATION in which the repetition frequency of PULSES is proportional to the amplitude of the modulating signal at any instant.

pulse generator a CIRCUIT that generates PULSES usually for testing pulse circuits. For comparatively long pulses, MULTIVIBRATORS can be used, but for pulses in microwave systems, pulse generators based on PULSE-FORMING LINES are more useful.

pulse-height analyzer a measuring instrument for received PULSES. The analyzer shows how many received pulses fall into each group of amplitude.

pulse modulation a form of modulation using a train of pulses to carry information. At MICROWAVE frequencies, conventional modulation of oscillators is almost impossible; and most oscillators are operated by pulsing. Only pulse modulation, therefore, can enable these frequencies to be used for communications. To take advantage of the large bandwidths available at microwave frequencies, MULTIPLEXER systems that allow more than one signal to modulate a pulse train are particularly useful. All pulse modulation systems operate on the basis that the pulse repetition rate will be much greater than the frequency of the signal being modulated, so that the pulses carry a sample of the signal.

pulse-position modulation a form of PULSE MODULATION using displacement of a pulse from its standard time position. This requires a clock signal to be available both at the transmitter and the receiver for detection of the deviation of the pulse from its correct timing.

pulse regeneration the restoration of a PULSE to its original specification. All circuits and transmission systems will distort the shape and timing of a pulse. A regeneration system restores the original shape and timing. The regeneration system often is a PULSE GENERATOR that is triggered by the imperfect pulse. All digital devices incorporate pulse regeneration, so that a degraded input pulse results in a perfect output pulse.

pulse repetition frequency (PRF) the rate in pulses per second at which PULSES are transmitted. KILO- and MEGA- are used with the unit of pps (pulses per second), so a rate of 5 kpps means 5 thousand pulses per second.

pulse shaper a CIRCUIT that alters PULSE shape. The INTEGRATOR and DIFFERENTIATOR circuits are pulse shapers, and a pulse re-generation circuit will consist of or contain a pulse shaper.

pulse spacing the time between pulses in a PULSE train.

pulse-time modulation (PTM) a general designation for any MODULATION system that affects timing rather than amplitude of pulses. See PULSE-DURATION MODULATION, PULSE-FREQUENCY MODULATION, PULSE-POSITION MODULATION.

pulse-width modulation see PULSE-DURATION MODULATION.

pumping 1. the alteration of a PARAMETER in a PARAMETRIC AMPLIFIER. **2.** the irregular rise and fall in noise level of the sound from a tape that has been recorded with an unsatisfactory noise-reduction system.

punch-through a breakdown of a region in a TRANSISTOR. Punch-through in a bipolar transistor means that the base region is destroyed by excessive collector-base voltage, so the base no longer controls the collector-emitter current. A similar effect occurs in field-effect transistors when current punches through the substrate layer, bypassing the gate.

puncture voltage the value of voltage that breaks down an insulator, particularly the DIELECTRIC of a capacitor.

purity the purity of color in a COLOR TELEVISION tube. Purity means that when a single color is displayed, it should appear as the same color with the same amount of saturation over the entire screen. The main cause of impure displays is accidental magnetization of metal components in or around the tube.

purity error any cause of impure color. See PURITY.

purple plague a common fault in very early TRANSISTORS. It was caused by the use of gold and aluminum together, which will form an alloy that diffuses into the SEMICONDUCTOR.

push-pull (of digital devices and amplifier output stages) having a pair of transistors or MOSFETS (see FIELD-EFFECT TRANSISTOR) that are used alternately to drive the output high (LOGIC ONE) or low (logic 0). See also OPEN-COLLECTOR DEVICE, TRI-STATE DEVICE.

pyroelectric (of a material) generating a voltage from a temperature gradient. Pyroelectric materials are used in fire alarms and other applications. Some types of plastic materials are strongly pyroelectric, as are several types of CRYSTALS.

Q

Q see Q-FACTOR.

Q band a MICROWAVE band in the wavelength range of approximately 6.5 mm to 8.3 mm, corresponding to the frequency range of 36 to 46 GHz.

Q or **Q-factor** the ratio of the reactance to the resistance of an electronic circuit or component.

Q meter an instrument for measuring the Q-FACTOR values of a RESONANT CIRCUIT.

quadrature the state of a VOLTAGE or CURRENT having a phase difference of 90 degrees to some reference phase. If, for example, the current wave has a phase angle of 90 degrees with respect to the voltage wave, the two are said to be in quadrature.

quadripole a network, particularly a balanced FILTER network, having four separate signal terminals.

quantization the SAMPLING of a signal to produce a number of values per second, an essential part of ANALOG/DIGITAL CONVERTER operation.

quantum a discrete amount of energy. A PHOTON is the quantum of electromagnetic energy and momentum absorbed or emitted in a single process by a charged particle.

quantum yield a measure of the efficiency of a PHOTOCATHODE equal to the average number of electrons produced for each photon of suitable energy.

quarter-wave match a LINE (1) that is a quarter of a wavelength long, used for matching two other lines.

quartz crystal the silicon-oxide crystal found naturally that is used as an electromechanical resonator. The crystal is cut to shape, and opposite sides are metalized (see METALLIZING). The crystal will then act as a TRANSDUCER of radio frequencies to mechanical vibrations and vice versa, and exhibit the behavior of a tuned circuit with very high Q-FACTOR. Such crystals are used to control the frequency of transmitters so as to be able to generate precise frequencies for timing, as in a digital watch or for SYNCHRONOUS DETECTION and in FILTERS with a narrow passband or stop band. See Fig. 75.

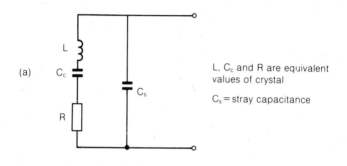

(a)

L, C_c and R are equivalent values of crystal

C_s = stray capacitance

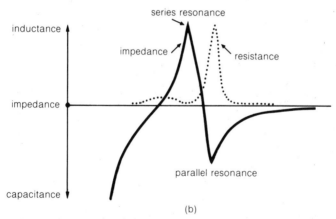

(b)

FIG. 75. **Quartz crystal.** (a) The equivalent circuit and (b) characteristics of a typical quartz crystal. The two resonant frequencies are usually close to each other in frequency. The equivalent Q-FACTOR is very large, of the order of 50,000.

quench deionize an ionized gas. The term can imply that ions in a gas-filled tube such as a GEIGER-MÜLLER TUBE have been made to recombine, so that the gas is no longer conductive.

quiescent current the steady current (BIAS current) that flows when no signal is applied to a circuit.

R

race hazard a problem in unsynchronized LOGIC GATE circuits. PULSES forming the inputs to a gate may be subjected to delays that are not the same for two different pulse paths. Because of this race, the two inputs to a gate may not arrive together, and the gate action is never achieved. Race hazards are avoided by SYNCHRONOUS operation.

radar a method of finding the distance and bearings of a metallic target, such as a ship, aircraft, missile, or possibly a nonmoving object. The term is derived from *ra*dio *d*etection *a*nd *r*anging, and the method measures the time between transmitting a wave pulse and receiving the reflected echo.

The simplest form of pulse radar generates a master pulse that is used to start a CATHODE-RAY TUBE timebase. The pulse is also used to modulate a MAGNETRON producing a short burst of waves. These MICROWAVES are transmitted from an antenna, using a parabolic reflector type to form a beam. When the waves strike a metallic target, a small proportion will be reflected. The returning wave is picked up in the same antenna, amplified, demodulated to form a voltage pulse, and displayed on the CRT screen. Because of the use of a timebase triggered by the initial pulse, the distance from the start of the sweep on the CRT screen represents the distance of the target.

A more common modern system uses a PLAN POSITION INDICATOR display, so both distance and bearings can be displayed. Lower power radar systems use continuous wave or FM systems, and the spread of moving objects can be measured using DOPPLER radar.

radar indicator a CATHODE-RAY TUBE display for a RADAR system.

radial-beam tube a vacuum tube intended for specialized purposes.

RADIATION PATTERN

radiation pattern a diagram that represents the effectiveness of an ANTENNA. For a transmitting antenna, it shows lines of constant field strength. For a receiver antenna, it shows lines of equal sensitivity. The usual pattern for a simple directional antenna is a figure eight, with unequal LOBES. The required direction is along the axis of the major lobe, and the rear lobe is often undesirable and represents wasted power in a transmitter.

radiation resistance see ANTENNA RESISTANCE.

radio 1. the technology of communication with electronically produced electromagnetic waves. **2.** any device that transmits or receives sound signals with these waves, and for a frequency in this range.

radio astronomy the process of locating and analyzing radio SIGNALS originating from stars or other distant bodies. This has become an important branch of astronomy, because many of the celestial sources of radio waves do not correspond to visible objects.

radio beacon see BEACON.

radio compass a form of radio direction finder that makes use of a DIRECTIONAL ANTENNA to locate the direction of a radio BEACON or other transmitter in a known location.

radio frequency (RF) a FREQUENCY in the accepted practical radio range. This covers the frequency range from about 100 kHz to about 300 GHz. At the lower end of the range, radiation is difficult unless extremely long antennas can be used. At the upper end of the range, electronic methods for generating signals can no longer be used.

radio-frequency heating the heating of objects by radio-frequency currents. The two main methods are induction heating (see EDDY CURRENT HEATERS) for metals and DIELECTRIC HEATING for insulators.

radio noise the unwanted NOISE signals that affect radio broadcasts. Noise can be natural or man-made. Natural noise consists of radiation from the movement of ions in the ionosphere and upper atmosphere along with signals directly from the sun and from other radio sources in space. Man-made noise includes interference from other transmitters, automobile ignition systems, fluorescent lights, electric machines, and other sources of radiation at radio frequencies. See also NOISE.

radio receiver a RECEIVER for radio signals that carry sound modulation. The sound signals for broadcast radio will be either amplitude modulated or frequency modulated onto the carrier. Military and other communications equipment will often use SINGLE-SIDE-BAND techniques.

Frequency modulation is used only for transmissions on VHF bands, because of the large bandwidth that is needed. The almost universal receiver is the SUPERHETERODYNE in which the received signal from the antenna is tuned and may be amplified in one stage, but then is converted to an INTERMEDIATE FREQUENCY for subsequent amplification. The demodulated IF signal is passed to the audio frequency stages, including any volume and tone controls, and then to the loudspeaker. At the demodulator, DC signals for gain and frequency control will be extracted. Special SYNCHRONOUS DETECTION techniques must be used if the transmission is of the single-sideband type.

FM receivers often make provision for STEREOPHONIC signal reception. Radio receivers can also be bought as TUNERS containing only the RF, IF, and first audio stages. These offer a higher quality of reception (of FM/stereo) and are intended for use with high-quality (hi-fi) amplifier and loudspeaker systems. High-quality reception of medium wave broadcasts is impossible because of the large number of transmitters that operate at closely spaced frequencies, plus pirate transmitters that in some areas make the use of medium wave reception difficult.

radiosonde an airborne measuring system for data transmission by radio from the upper atmosphere. Radiosondes are used by meteorologists. In these systems, balloons carry small transmitters and transducers to the required altitude. The transmitters then send back data on atmospheric conditions until they ultimately are lost.

radiotelescope an instrument for receiving radio signals from space. The main antenna types are parabolic dishes or arrays. The rest of the receiving equipment follows conventional radio receiver designs.

radio wave see CARRIER.

RADIO WINDOW

radio window the range of FREQUENCIES not reflected by the IONOSPHERE and, therefore, useful in RADIO ASTRONOMY. The range is roughly from 15 MHz (20 m wavelength) to about 50 GHz (6 mm wavelength).

RAM see RANDOM-ACCESS MEMORY.

ramp a steadily rising or falling voltage, such as the linear SWEEP portion of a timebase.

random-access memory (RAM) a type of electronic MEMORY that enables data to be read to or written from any location independently of others. A *serial-access memory*, by contrast, allows access to a unit of data only by shifting other units out. Random access permits only the desired unit to be affected.

This is done by allocating an address number to each unit of data. A set of address connections to the memory can be used to convey an address number, so that when an address number is placed in binary form on the lines, the correct data unit is made available. Another connection is used to select between reading or writing. Compare READ-ONLY MEMORY.

range the distance between TRANSMITTER and RECEIVER. The term is often used to denote the *maximum range*, the greatest distance at which reliable communication can be established.

range tracking a form of RADAR system that enables a moving target to be followed, but reduces other echoes. The echo from the target is used to generate a gating signal for the target echo.

raster a scanning pattern on a TV screen. Conventional TV uses a SCAN pattern of horizontal lines that slope slightly. Each screen line starts on the left-hand side of the screen and sweeps across to the right, moving downward slightly as it does so.

The line pattern is produced by using a horizontal TIMEBASE waveform, and the vertical movement from a vertical (FIELD) timebase. These timebase signals are applied to sets of DEFLECTION COILS arranged around the neck of the tube. In this way, a pattern of parallel lines is traced out on the screen. In a complete picture, two interlaced sets of lines are used. This means that the first set of lines, the even-numbered lines, will be spaced so as to occupy the vertical dimension of the screen. The second set, the odd-numbered lines, will then occupy the spaces between the first lines. This pattern is used to satisfy the contradictory requirements of good vertical resolution, lack of flicker, and narrow bandwidth.

rating a specification of maximum operating conditions. Ratings of voltage, current, frequency, temperature, etc. for a component or assembly are determined by the manufacturer.

ratio adjustor see TAP CHANGER.

ratio detector a form of FM DEMODULATOR. The ratio detector is useful only if there is no trace of amplitude modulation in the signal, and the detector must be preceded by a limiter stage to ensure this. See also FREQUENCY DISCRIMINATOR.

rat-race junction see HYBRID JUNCTION.

reactance the ratio of SIGNAL voltage to signal current in an ideal CAPACITANCE or INDUCTANCE, that is, one with no measurable series resistance and infinite parallel resistance. The value of reactance depends on the signal frequency as well as on the value of capacitance or inductance. A pure reactance also causes a 90-degree phase shift between voltage and current. For a capacitor, the phase of current leads the phase of voltage, and for an inductor, the phase of current lags the phase of voltage. See Fig. 76. See also IMPEDANCE, RESISTANCE.

reactance chart a type of NOMOGRAM used to show reactance for different values of capacitance, inductance, and frequency.

reactance transformer an IMPEDANCE-matching circuit that uses REACTANCE.

reactivation the reviving of electron EMISSION from the cathode or filament of a vacuum tube. This usually involves running at higher than normal temperature for some time, often with grid current being passed.

reactive component or **wattless component** the product of signal VOLTAGE and CURRENT when the PHASE DIFFERENCE is 90 degrees. This quantity does not represent dissipated power, since only the product of the in-phase component's voltage and current in phase gives the true figure of power dissipated.

reactive current or **idle current** the component of current whose phase angle is at 90 degrees to the phase of voltage.

reactive load any LOAD that causes a phase angle between the current and the voltage of a signal wave applied to it, that is, a load that contains REACTANCE.

REACTIVE VOLTAGE

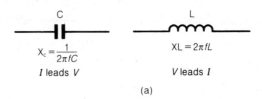

$$X_c = \frac{1}{2\pi fC}$$

I leads *V*

$$XL = 2\pi fL$$

V leads *I*

(a)

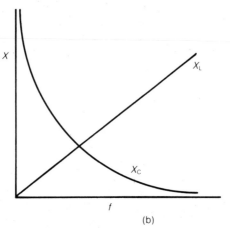

(b)

FIG. 76. **Reactance.** The reactance of capacitors and inductors, showing (a) value formulas, (b) the general reactance-frequency graph. The mnemonic for phase relationship is CIVIL, that is, Capacitor, I before V; V before I in L (inductor).

reactive voltage the component of voltage whose phase angle is at 90 degrees to the phase of current.

reactive volt-amperes the product of voltage and current when a phase angle exists. The dissipated power, which is less than this value, is found by multiplying the reactive volt-amperes figure by the power factor, the cosine of the phase angle.

reactor a COMPONENT that is REACTIVE such as a capacitor or inductor.

read copy information that is stored, particularly in a computing storage system. A nondestructive read process will allow the copy to be made in the form of a voltage on a line without any effect on the stored information. Compare WRITE. A DESTRUCTIVE READ process will destroy the stored information in the course of copying it. For example, the stored charge image in a TV camera tube is read by measuring the electron current needed to discharge it.

read-only memory (ROM) a type of MEMORY for microprocessor or computer use. The usual form of ROM permits random access to any unit, but for purposes of reading only. This type of memory is NONVOLATILE and is used for all programs that must be present from the instant of switching on. Compare RANDOM-ACCESS MEMORY.

read-out pulse a PULSE applied to a magnetic core MEMORY system that enables a selected part of the memory to be read.

read-write head the TRANSDUCER of a tape or magnetic disk recording system. See MAGNETIC RECORDING.

receiver a device that processes radio signals from an ANTENNA into final form—audio, video, digital code, etc. Compare TRANSMITTER.

reception the interaction between an electromagnetic wave and any form of antenna that results in a signal being received. Compare TRANSMISSION.

reciprocal theorem a theorem relating to networks. Imagine two points, A and B, in a linear network. If a signal voltage V at point A causes a signal current I at point B, then a voltage V at point B will generate a current I at point A.

recombination the removal of CHARGE carriers from a semiconductor or an ionized gas. Recombination can occur when an ELECTRON and a HOLE meet and release energy, so that the electron is no longer mobile and the hole no longer exists. Both electrons and holes can also be trapped by impurity atoms.

record 1. a permanent store of data. 2. a phonograph record. See RECORDING OF SOUND.

recorder an item of recording equipment, for example, a tape recorder, a cassette recorder, or a multichannel tape recorder for recording data. See also CHART RECORDER.

recording the storing of data in permanent form.

recording of sound the use of a system of TRANSDUCERS to convert sound into electrical signals that can be recorded. The main systems are mechanical analog, mechanical digital, magnetic, and optical.

The mechanical analog uses the vibration of a stylus to modulate the shape of a groove that is cut into a phonograph DISK. The mechanical digital uses an ANALOG/DIGITAL CONVERTER and records the digital signals by pitting the surface of a disk with a laser beam. Storage density is very high, and the system permits the highest standard of SOUND REPRODUCTION that has been achieved to date, the COMPACT DISK system. The magnetic systems use either analog or digital recording on magnetic tape. The optical system is used only for film soundtracks, in which the width or density of a strip of film is modulated by the audio signals. Optical systems suffer from poor signal-to-noise ratios, and noise-reducing techniques such as the DOLBY SYSTEM are very useful. Tape systems that use slow-moving, narrow tapes also suffer from noise problems, and the same solutions are used. See also DBX.

recording of video the use of a system for recording and replaying video signals, usually on tape (see VIDEO CASSETTE). The recording of video signals on tape requires processing of the signals to reduce their bandwidth. The video-recording heads are moved to sweep along and across the tape, so that the speed of the tape can be comparatively low while permitting the speed of the head relative to the tape to be high. The tape is considerably wider than conventional audio-recording tape.

The sound track is separately recorded, usually on a strip that runs longitudinally, though on some modern machines much higher-quality sound reproduction is obtained by using an additional track swept with the video head. Video recording of a much higher standard is also possible on disks, using the same laser read and write techniques used for compact (audio) disks.

Read-write laser disks are under development. See Fig. 77.

rectifier a COMPONENT such as any type of DIODE that passes current easily in one direction only. A rectifier, used in a half-wave or full-wave circuit, will convert AC into UNIDIRECTIONAL CURRENT. This can be used to charge a RESERVOIR CAPACITOR in order to obtain smooth DC.

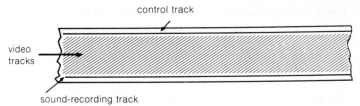

control track

video tracks

sound-recording track

FIG. 77. **Recording of video.** The arrangement of the recording head tracks on a typical videotape. The soundtrack is at one edge, and a control track is at the other edge. The video recording tracks run diagonally across the tape.

rectifier instrument a measuring instrument that can be used with AC because it contains a RECTIFIER. The rectifier-bridge circuit is normally used.

rectifier voltmeter a VOLTMETER that can measure sine wave AC values (see ROOT MEAN SQUARE) because it incorporates a RECTIFIER. The rectifier may be switchable, so that the voltmeter can be used for either DC or AC.

rectifying contact a CONTACT between two materials that has a rectifying action. The difference between the forward and the reverse resistance values may be small, but the presence of the rectifying contact can cause nonlinear effects in a signal.

red gun one gun of a COLOR CRT. The red gun is the gun whose electrons will pass through the APERTURE GRILLE to strike only the parts of the PHOSPHOR that glow red.

redundancy a deliberate or natural duplication. Redundancy in a circuit means that additional components are provided that will take over from other components in the event that those other components fail. This idea can extend to complete systems, such as on-board computers for space or aircraft systems.

Redundancy in a signal means information that is not needed. In a TV signal, for example, over 90% of the data transmitted on a frame is also transmitted on the next frame, unless there is a total change. Some types of narrow bandwidth systems operate by removing redundancy from signals.

reflected current the signal current reflected from a faulty termination of a TRANSMISSION LINE.

reflected impedance an IMPEDANCE whose effect can be measured in a circuit, but is connected to that circuit by means of a transformer, line, or other coupling. The reflected impedance of the secondary of a transformer is the amount of impedance measured at the primary that is due to the presence of the secondary load.

reflected power the POWER not absorbed by a LOAD and is returned to the generator. This applies typically to a TRANSMISSION-LINE system that is not correctly matched. In a microwave system, reflected power can cause overheating and failure of the microwave generator.

reflected wave 1. a wave reflected in a TRAVELING-WAVE TUBE. 2. Another name for SKY WAVE.

reflection a reversal in the direction of travel and phase of a WAVE usually because of striking a metal or other conducting surface, or from a discontinuity in a transmission line or other medium.

reflection coefficient the ratio of reflected current to forward current in a line. The two currents in general will not have the same PHASE.

reflection error an error in RADAR or RADAR COMPASS systems caused by false REFLECTIONS of signals.

reflection factor a measure of MATCHING in a TRANSMISSION LINE. The reflection factor is the ratio of the current sent to a reflecting load to the current that would be sent to a perfectly matched load. The factor is often converted into DECIBELS.

reflector an element of a DIRECTIONAL ANTENNA. The reflector is placed behind the active portion of the antenna, which is usually a DIPOLE. Its action is to reflect signal back to the dipole and so greatly reduce the size of the rear LOBE. The reflector can be made of sheet metal or even of wire mesh. The length of the reflector is always greater than that of the dipole, usually by a factor of about 10%, and the spacing between reflector and active element is of the order of a quarter of a wavelength.

reflex circuit a circuit formerly used as an AMPLIFIER for a pair of signals at different frequencies. It was used as a way of economizing on tubes in early radio receivers.

reflex klystron a form of KLYSTRON in which the beam is reflected back after bunching. The effect is to make the device into a MI-

CROWAVE oscillator whose frequency is determined by the cavity dimensions and by the voltage applied to the reflecting electrode. Reflex klystrons once were the only devices available for use as local oscillators in microwave receivers. They have been almost completely superseded for low-power applications by GUNN DIODES.

reform electrolytic apply steady POTENTIAL DIFFERENCE across an ELECTROLYTIC CAPACITOR until the leakage current is at a minimum. Reforming is sometimes needed after an electrolytic capacitor has been stored unused for a long period, or after exposure to cold conditions. The reforming voltage reestablishes the film of aluminum oxide that acts as the insulator.

refresh regenerate the condition of a signal. A refresh operation has to be used in storage devices that use destructive readout. The term is applied today mainly to DYNAMIC RAM which uses MOS capacitors to store data in the form of charges. Because of unavoidable leakage from these capacitors, the memory must be refreshed by recharging at intervals of, typically, one millisecond.

regeneration see POSITIVE FEEDBACK.

regenerative receiver an early type of RECEIVER that achieved very high GAIN and SELECTIVITY by using RF positive feedback. See also SUPER-REGENERATION.

register an assembly of connected FLIP-FLOPS in which loading, storage, or serial shifting (see SERIAL MEMORY) can be performed. Registers are used in microprocessors and in the central processing units of large computers. Registers are used to store data, in particular the results of arithmetic and logic actions on data.

regulator a CIRCUIT that can maintain a constant value of either VOLTAGE or CURRENT under a wide range of conditions.

reignition voltage the VOLTAGE needed to start CURRENT flowing in a gas-filled device. If a trace of IONIZATION remains from previous current flow, the reignition voltage will be lower than the usual cold striking voltage.

rejection band or **stop band** the range of frequencies a FILTER will reject. Compare PASSBAND.

rejector a CIRCUIT with very high IMPEDANCE at one frequency. The term usually denotes a PARALLEL RESONANT CIRCUIT used as a band-stop filter.

relative permeability the ratio of the permeability of a material relative to the PERMEABILITY OF FREE SPACE. Symbol: μ_r.

relative permittivity the ratio of the permittivity of a material relative to the PERMITTIVITY OF FREE SPACE. Symbol: ϵ_r. The relative permittivity value is equal to the value of absolute permittivity of the material divided by the permittivity of free space.

relaxation oscillator or **aperiodic oscillator** a form of untuned oscillator in which the active devices pass current only for brief intervals. MULTIVIBRATORS, UNIJUNCTION oscillators, and BLOCKING OSCILLATORS are all relaxation oscillators. The output waveform of such an oscillator is aperiodic and consists of steep-sided waveforms with comparatively long periods of steady voltage, either at zero or supply voltage.

relaxation time the recovery time for a CAPACITOR following a sudden change of charge.

relay an electromechanical SWITCH. A current flowing through a coil causes a magnetic field that attracts an armature. The armature moves, opening or closing switch contacts that are mechanically connected to it. Many uses of relays have now been taken over by purely electronic components, such as THYRISTORS and TRIACS. In some cases, however, wiring regulations still insist on the use of mechanical relays.

relay driver a CIRCUIT that provides current to the coil of a RELAY. When a TRANSISTOR is used as a relay driver, a diode must be connected to absorb the pulse of back-emf that will be generated when current is switched off.

reliability a measure of confidence in the ability of a DEVICE to go on working. Unlike mechanical devices, the reliability of electronic equipment is too high to measure in terms of probability. A more useful measure is the inverse, FAILURE RATE.

reluctance a magnetic circuit quantity comparable with RESISTANCE. See FLUX, MAGNETOMOTIVE FORCE.

remanence or **residual magnetism** the retained MAGNETISM of a magnetic material. The remanence is the FLUX DENSITY remaining in material that has been magnetized to saturation, after which the magnetizing field has been removed. See also MAGNETIC HYSTERESIS.

remote control the control over a DEVICE at a distance, by electrical signals, radio, or other electronic signals.

repeater a form of signal BOOSTER. A repeater in a telephone line amplifies the audio signals to compensate for attenuation in the lines. Repeaters can be made that operate in either direction, or restore the original waveform of signals such as pulses.

reset restore anything to its original setting. As applied to a register, reset means to clear, making each output logic 0.

reservoir capacitor an essential part of AC-to-DC conversion. The reservoir capacitor is charged by a conducting rectifier to the peak voltage of the waveform. When the input voltage to the rectifier drops, the reservoir capacitor supplies current to the load until the rectifier can conduct again. See Fig. 78.

residual charge the small amount of CHARGE remaining in a CAPACITOR after a spark discharge.

residual current a CURRENT that continues to flow for a short time, measured in nanoseconds, after voltage has been reduced to zero. The effect is caused by the momentum of the CHARGE CARRIERS.

residual magnetism see REMANENCE.

resistance the ratio of DC voltage to DC current for a conductor. Symbol: R; unit: ohm. The inverse of resistance is CONDUCTANCE. See also OHM'S LAW.

resistance-capacitance (R-C) coupling a method of transferring a signal from one amplifier STAGE to another. A resistor is used as a load in the output stage of one device, and a capacitor is used to transfer AC signal without affecting the DC level of the next stage.

resistance pressure gauge a form of pressure gauge that measures pressure by its effect on a resistor.

resistance strain gauge a STRAIN GAUGE that uses a wire element. The wire is bonded to the material under test, and changes in strain cause variations in the wire length and cross section that alter the wire's resistance. Semiconductor strain gauges are easier to use because of their much greater sensitivity, but are more costly.

resistance thermometer an instrument for measuring tempera-

RESISTANCE WIRE

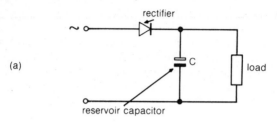

(a)

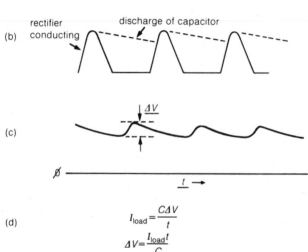

(b)

(c)

(d)

$$I_{\text{load}} = \frac{C \Delta V}{t}$$

$$\Delta V = \frac{I_{\text{load}} t}{C}$$

FIG. 78. **Reservoir capacitor.** (a) A half-wave power supply circuit. (b) The action of the reservoir capacitor. (c) The supply waveform when the load draws a substantial current. (d) The approximate formula for ripple voltage.

ture electrically. The resistance of any metallic conductor changes with temperature, becoming greater as temperature rises. Resistance thermometers use wire coils, made of nickel alloy or platinum, as sensors, and the resistance is usually measured in a BRIDGE circuit, using COMPENSATING LEADS.

resistance wire a wire made of high-RESISTIVITY material. Typical materials are nickel-chromium alloys, such as NICHROME or CONSTANTAN.

resistive component the COMPONENT of an IMPEDANCE for which voltage is in phase with current.

resistivity a quantity that measures the amount of RESISTANCE due to a material and independent of the dimensions of a sample of the material. Symbol: ρ; unit: ohm-meter. The figure of resistivity for a material is independent of shape or size and can be used to calculate the resistance of any specimen of that material. For a sample with area of cross section A and length d, and with resistance R, the resistivity is given by RA/d. Compare CONDUCTIVITY.

resistor a COMPONENT used for its RESISTANCE. In the past, most resistors were manufactured from carbon composition, a baked mixture of graphite and clay. These have been almost completely superseded by carbon or metal FILM RESISTORS. WIRE-WOUND RESISTORS are used for comparatively low values of resistance where precise value is important, or for high dissipation. They are unsuitable for RF use because of their reactance. See also VARIABLE RESISTOR.

resistor-transistor logic (RTL) an obsolete system of logic circuitry for INTEGRATED CIRCUIT use. These circuits use integrated resistors and transistors to form logic GATES (2).

resolution or **definition** the ability of a display system to show fine detail, particularly cathode-ray and LCD displays.

resolution of potentiometer a measure of the ability of a variable POTENTIOMETER to produce a small voltage change by a small adjustment. When wire-wound potentiometers are used, the smallest percentage change in resistance is given by the change from one turn of the winding to the next. For continuous metal-film potentiometers, this restriction does not apply, and greater resolution can be obtained.

resonance the maintenance of OSCILLATION with minimum driving signal. A RESONANT CIRCUIT will show little response to signals until a signal close to the natural resonant frequency is used. At the resonant frequency, maximum current flows in the resonant circuit components for minimum input signal. There is usually current or voltage magnification, so the circuit acts as a form of selective amplifier for one frequency.

resonance bridge a form of BRIDGE circuit in which one arm consists of a series-tuned circuit. The bridge can be balanced only at the resonant frequency of the tuned circuit. From the balance conditions, the resistance of the tuned circuit can be found.

resonant cavity a closed space surrounded by metal that acts as a RESONANT CIRCUIT for MICROWAVES.

resonant circuit or **tuned circuit** a circuit that will resonate at some frequency. Lumped resonant circuits use a capacitor and an inductor, but resonance can also occur in a quartz crystal, a transmission line, and a cavity.

resonant frequency see NATURAL FREQUENCY.

resonant line a TRANSMISSION LINE whose inductance and capacitance resonate at the operating frequency.

rest mass the mass of a body as measured when the body is at rest relative to an observer.

retentivity the quantity that measures the amount of retained magnetism of a magnetic material after an applied field has been removed. The units are webers per square meter. See also MAGNETIC HYSTERESIS, COERCIVITY.

reverberation unit (reverb. unit) a form of ULTRASONIC DELAY LINE used to provide artificial echo in a sound recording.

reverse AGC a form of AUTOMATIC GAIN CONTROL that uses NEGATIVE BIAS. Reverse AGC is used when the gain of the amplifier stages is reduced by reducing their operating current. Compare FORWARD AGC.

reverse bias a BIAS that opposes the flow of current in a device. Compare FORWARD BIAS.

reverse direction the direction of high resistance to current flow through a device.

RF see RADIO FREQUENCY.

RF heating eddy current heating (see EDDY CURRENT HEATER) or DIELECTRIC HEATING using radio frequencies.

rhombic antenna a four-sided wire structure that acts as a DIRECTIONAL ANTENNA.

ribbon microphone or **induction microphone** a form of dynamic-induction MICROPHONE. A thin metal ribbon is stretched between the poles of a high-flux magnet. Vibration of the ribbon by sound waves causes an induced signal voltage, which can be amplified. The microphone uses no separate diaphragm, thus reducing unwanted resonances, and is particularly suitable for high-quality work. The output is very small, but because of the low impedance it is not difficult to avoid interference pickup. See also VELOCITY MICROPHONE.

right-hand rule see FLEMING'S RULES.

ring counter a form of digital counter that uses FLIP-FLOPS connected in a circular arrangement, so that the output of the last flip-flop's is connected to the input of the first flip-flop.

ringing 1. the oscillation of a resonant circuit caused by a pulse signal (SHOCK EXCITATION). 2. unwanted oscillations appearing in a pulse.

ripple the proportion of AC remaining in a steady voltage from a power supply. The ripple is due to the fluctuations of voltage across the RESERVOIR CAPACITOR as it is charged and discharged. For a full-wave rectifier system, the ripple frequency is twice the supply frequency. For a half-wave circuit, the ripple will be at supply frequency. The ripple waveform is generally a sawtooth rather than a sine wave.

ripple counter a form of digital counter consisting of a chain of FLIP-FLOPS. The input pulses are used to switch the first flip-flop, and the output of this flip-flop is used to toggle the next in the chain. The number of pulses is shown in terms of the digital number stored by the outputs of each flip-flop. Because each flip-flop input is the output of the previous stage, increasing the count by one will cause a successive series of charges to ripple through the device.

ripple factor the percentage of RIPPLE in a DC supply. This is expressed as the percentage of ROOT MEAN SQUARE ripple, assuming that the ripple is a sine wave, which it may not be.

ripple filter an integrating or smoothing circuit that removes RIPPLE. The circuit is a simple form of LOW-PASS FILTER using series inductance or resistance and parallel capacitance.

ripple frequency the frequency of RIPPLE in terms of supply frequency. If a bridge rectifier is used, ripple occurs at twice the supply frequency. If a half-wave rectifier is used, ripple occurs at supply frequency.

rise time the duration of a pulse LEADING EDGE. This is usually taken as the time for the voltage to change from 10% to 90% of the pulse peak voltage.

RMS see ROOT MEAN SQUARE.

robotics the study of machines with some form of intelligence. The term robot generally denotes any machine that has been programmed to carry out a set of repetitive mechanical actions with a minimum of human intervention.

ROM *acronym for* read-only memory, a form of memory for computers. The ROM operates with random access, but can only be read, not written. This makes it suitable for programs and other data that have to be available at the instant of switching on. ROMs for modern systems are manufactured by integrated circuit masking, making permanent connections within an IC to logic 0 or 1, along with memory addressing circuits so that the state of each connection can be read. Semipermanent programs can be stored in PROM form. See also EPROM, EAROM.

root mean square (RMS) 1. the square root of the average of the squares of a set of numbers or quantities. 2. the root of the mean value of the square of wave amplitude. The root is used to find the DC equivalent of an AC quantity. The principle is that power can be calculated correctly for the AC circuit using RMS values, provided that PHASE SHIFTS are correctly taken into account. For a sinusoid, the RMS value is equal to peak value divided by $\sqrt{2}$. A different factor must be used for other waveforms. See also TRUE RMS.

rotary encoder an ANALOG/DIGITAL CONVERTER for a rotating shaft, which often uses an optical source/sensor arrangement and a slotted disk.

rotating-anode tube a form of X-RAY TUBE in which the ANODE can be rotated to avoid overheating.

rotation or **endaround shift** a serial memory or REGISTER action in which the most significant bit from the end of a memory or register is loaded in at the least significant end, the remaining bits being shifted one place to make room.

rotator a portion of a WAVEGUIDE that changes the plane of polarization of the waves.

rotor the rotating part of a motor or generator. Compare STATOR.

RS232 an internationally agreed standard for serial data transmission. It specifies high and low voltage levels, timing, and control.

R-S flip-flop the simplest type of FLIP-FLOP constructed from AND and NOT gates. An output is obtained if one input at the two terminals, R and S, is at LOGIC ONE. The output is then stored for as long as both inputs are held at logic zero. The R-S flip-flop was superseded by the J-K FLIP-FLOP except for the purpose of switch debouncing. See Fig. 79.

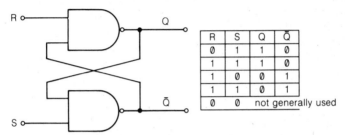

R	S	Q	Q̄
0	1	1	0
1	1	1	0
1	0	0	1
1	1	0	1
0	0	not generally used	

FIG. 79. **R-S flip-flop.** An R-S flip-flop constructed from NAND gates. The 1,1 input is a holding input. The circuit can also be constructed from NOR gates, and the holding input is the 0,0, and the forbidden input is 1,1.

rumble a low-frequency NOISE. The term usually denotes low-frequency audio signals caused by mechanical vibrations of the motor of a phonograph.

runaway a loss of control, for example, THERMAL RUNAWAY the loss of control of temperature. If the temperature of a transistor increases to such an extent that the base current is no longer able to control collector current, the current and temperature will increase until the junctions are destroyed. Thermal runaway is rare with modern silicon transistors.

S

saddle-shaped field the shape of the ELECTROSTATIC FIELD required of an ELECTRON LENS to focus a beam of electrons. The field can be obtained from a set of three electrodes, with the central electrode at a lower potential than the other two. See Fig. 80.

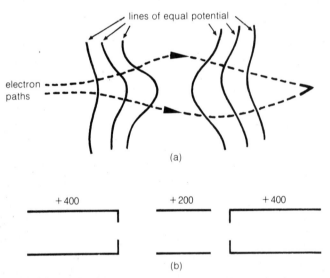

(a)

(b)

FIG. 80. **Saddle-shaped field.** (a) The saddle-shaped field of lines of equal potential, and the electron paths (dotted) in such a field. (b) A cross section of a typical electrode structure, which will produce a field shaped as shown in (a).

sag the decay of amplitude of a square PULSE. If the input to a circuit is a perfect square pulse, sag on the output pulse is caused by poor low-frequency response. The amount of sag is given as a percentage of peak amplitude.

sampling the making of measurements at intervals. Sampling is an essential part of ANALOG/DIGITAL CONVERTER operation, and the conversion will be useful only if the sampling rate is high enough. A minimum sampling rate of twice the sample frequency is often cited, but in practice a minimum of five or ten times is used. See also SHANNON'S SAMPLING THEOREM.

saturable reactor or **transductor** an INDUCTOR whose inductance value can be considerably changed by passing DC through another coil wound on the same core. See also MAGNETIC AMPLIFIER.

saturated mode the constant-current operation of a FIELD-EFFECT TRANSISTOR. A FET is in the saturated mode when the DRAIN voltage is kept at a level at which only the GATE voltage affects the drain current.

saturation 1. a limiting state of any variable quantity normally controlled by changing another quantity. A magnetic material is saturated when the magnetic flux density of a material is almost unaffected by changes in the magnetizing field. **2.** See HUE.

saturation current a state of maximum current, unaffected by a rise of voltage. See SATURATED MODE.

saturation resistance the RESISTANCE between COLLECTOR and EMITTER (1) of a saturated bipolar TRANSISTOR. SATURATION occurs because the resistance falls to such a low value that the current is determined by the load resistor rather than by the base current.

saturation signal a SIGNAL that overloads a radar RECEIVER.

saturation voltage the VOLTAGE between COLLECTOR and EMITTER (1) of a saturated bipolar TRANSISTOR. For a silicon transistor, this is typically less than 200 mV.

SAW see SURFACE ACOUSTIC WAVE.

sawtooth oscillator an OSCILLATOR whose output is a voltage or current wave of sawtooth shape. The sawtooth oscillator is a form of RELAXATION OSCILLATOR.

sawtooth waveform the output WAVEFORM from a SAWTOOTH

OSCILLATOR. An ideal sawtooth waveform consists of the linear SWEEP section, a brief FLYBACK and a waiting period between the end of the flyback and the start of the next sweep.

S band a MICROWAVE band, with wavelength range from 5.77 to 19.3 cm, corresponding to the frequency range of 1.55 to 5.2 GHz.

scale of integration a description of the number of DEVICES obtainable on a single CHIP. The integration scales range from SSI (small scale), through MSIand LSI to VLSI (very large scale integration).

scaler a form of PULSE-count divider. A scaler gives a pulse output after a fixed number of input pulses have been received. Most scalers are decade—ten pulses in for one pulse out; or binary—two, four, eight, or sixteen pulses in for one pulse out.

scan the scanning movement of an ELECTRON BEAM across a target or screen. Television requires both a horizontal and a simultaneous vertical scan. See also RASTER.

scanner a detector system for radioactive material. The term usually denotes a detector of radio tracers in the body, allowing a map to be made of suspect tissue.

scanning see SCAN.

scanning yoke an arrangement of DEFLECTION COILS for a TV or radar CATHODE-RAY TUBE.

scattering loss the loss of energy from an electron or light or radio BEAM by irregular reflections. Loss by scattering is particularly noticeable when the wavelength of the radiated energy is close to the dimensions of scattering particles, for example, the effect of fog on light. The use of metal strips, called *chaff* or *window*, to confuse radar is an example of deliberate scattering of electromagnetic waves. An electron beam is scattered when it passes through matter.

schematic see CIRCUIT DIAGRAM.

Schmitt trigger a form of level-triggered BISTABLE CIRCUIT. The output of a Schmitt trigger remains low until the input voltage reaches a critical level, when it switches to high. The input voltage must then be lowered to a different and lower critical level in order to switch the output voltage to low again. The Schmitt trigger is a particularly useful circuit for PULSE REGENERATION. See Fig. 81.

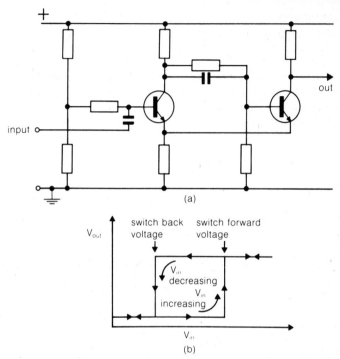

FIG. 81. **Schmitt trigger.** (a) The Schmitt trigger circuit. (b) Its CHAR-
ACTERISTIC. The characteristic exhibits hysteresis that can be very
useful, particularly for digital circuits. Schmitt triggers are available in
integrated circuit form.

Schottky diode or **hot-carrier diode** a DIODE that uses a metal-
semiconductor JUNCTION usually aluminum-silicon. The diode has
a very low forward voltage for conduction, fast switching speed,
and fast cutoff on reverse bias. Schottky diodes are incorporated
into a range of digital logic integrated circuits, the transistor-transistor logic series, in order to achieve faster switching with lower
values of current.

Schottky noise or **shot noise** a form of NOISE caused by the
random emission of electrons, particularly in vacuum devices.

279

scintillation 1. the flash of light in some types of crystals when struck by an ionizing particle. **2.** the random variation of amplitude in a radio signal that has been transmitted over a long distance.

scintillation counter a detector and counter of IONIZING RADIATION. The active elements consist of a scintillating crystal and a PHOTOMULTIPLIER. The flash of light produced by the CRYSTAL for each ionizing particle is detected by the photomultiplier, and the output pulse from the photomultiplier is amplified and used to operate a SCALER or COUNTER. The color of light from the scintillation counter depends on the energy of the particle that produced it, and the pulse amplitude from the photomultiplier is also affected by the energy of the radiation. This makes it possible to analyze the output of a scintillation counter in more detail than is possible when a GEIGER-MÜLLER TUBE detector is used.

SCR see THYRISTOR.

scrambler a DEVICE for making telephone communication secure. The signal is treated, for example, by inverting the frequency bands, so it makes no sense to listeners who do not have unscramblers for decoding.

screen the endplate of a CATHODE-RAY TUBE or the metal shield against electric or magnetic fields. See FARADAY CAGE, MAGNETIC SCREENING.

screen burn see ION BURN.

screened pair two insulated signal-carrying wires surrounded by a conductive braid. A screened pair is commonly used for stereo audio signals, with one wire for each channel and the braid used for the ground connection.

screening the encasing of a CIRCUIT with metal to reduce the effect of MAGNETIC or ELECTRIC FIELDS. The screening is usually against external fields, but it can also be used to prevent internal fields from escaping. For electrostatic screening, the metal should be a good conductor and grounded. For magnetic screening, the metal must be of high PERMEABILITY and in the form of a complete loop.

SCS see SILICON-CONTROLLED SWITCH.

search coil a small COIL used along with a ballistic galvanometer as a method of measuring flux. This method of measuring flux has been largely superseded by the use of HALL-EFFECT devices.

SECAM *acronym for* Système Electronique Couleur Avec Mémoire, the name of the French COLOR TELEVISION transmission system, which differs in several respects from the NTSC and PAL systems. In the SECAM system, the color difference signals, **R-Y** and **B-Y** are sent alternately. The signals are frequency modulated onto the SUBCARRIER and a delay line is used at the receiver to enable both sets of signals to be available.

The use of sequential signals effectively halves the vertical resolution, as the summing of signals does in the PAL system, and the use of frequency modulation causes marginally more visible dot-patterning on monochrome receivers than the PAL system. PREEMPHASIS is used to reduce the effect of the frequency variation of the subcarrier.

secondary cell a CELL whose chemical action is reversible. The cell will provide electric energy until the chemical action is almost complete, at which point the action can be reversed by connecting the cell to a source of POTENTIAL DIFFERENCE higher than the normal output potential difference of the cell. This recharging action can be continued until the chemicals in the cell have been completely restored to their original state, enabling the cell to be used again. See also LEAD-ACID CELL.

secondary electron an ELECTRON emitted from a material in a SECONDARY EMISSION process. Compare PRIMARY ELECTRON.

secondary emission the EMISSION of electrons from a material because of electron bombardment of PRIMARY ELECTRONS. The effect occurs only for a limited range of accelerating voltage of the primary electron beam, between the crossover voltages. Below the first crossover voltage, usually 30 to 100 volts, primary electrons land or are reflected. Above the second crossover voltage, in the region of 2,500 to 4,000 volts, primary electrons penetrate the material and are retained. In the secondary emitting region, more electrons are emitted at low velocities than land. If the material is an insulator, it will become positively charged as a result of the secondary emission. See Fig. 82.

secondary radar a radar system in which reflections are not used. See IFF, INTERROGATING SIGNAL, TRANSPONDER.

SECONDARY STANDARD

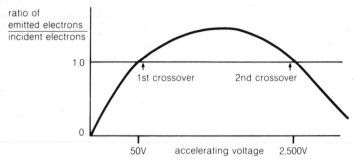

FIG. 82. **Secondary emission.** A typical secondary-emission graph for a material, showing the voltages of the first and second crossovers. Between these crossover points, more electrons are leaving the surface than land on it. If the substance is an insulator, its voltage will stabilize at cathode potential for bombardment at a voltage lower than that of the first crossover. For a bombarding voltage of a value between the values of the crossover points, the surface will stabilize at the anode voltage. For bombardment using a voltage greater than that of the second crossover, the insulator potential will stabilize at the second crossover potential. This is one reason for using an aluminized screen for a cathode-ray tube. In the absence of aluminizing, the screen surface is at the second crossover potential rather than at the anode potential.

secondary standard a standard of measurement that can be used for calibrating instruments. A secondary standard, sometimes confusingly known as a *substandard*, is calibrated from a PRIMARY STANDARD. The secondary standard then becomes the practical implementation of the standard.

secondary voltage the AC voltage across the SECONDARY WINDING of a TRANSFORMER.

secondary winding the TRANSFORMER winding across which an output of AC is taken. In the older core type of transformer, the secondary winding was the outside winding, which was wound over the primary winding. In modern toroidal and shell-construction transformers, the secondary and PRIMARY WINDINGS are interchangeable, and the only difference is in function. See also TERTIARY WINDING.

Seebeck effect see THERMOELECTRIC EFFECT.

seesaw circuit an INVERTER with unity gain. The seesaw circuit uses 100% NEGATIVE FEEDBACK in an INVERTING AMPLIFIER and its main application is for inverting a signal to apply to a balanced amplifier. See Fig. 83.

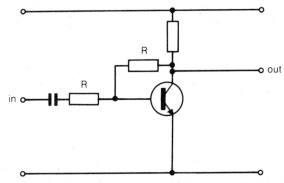

FIG. 83. **Seesaw circuit.** The seesaw circuit, or PARAPHASE AMPLI-
FIER, in a direct-coupled form. The resistors marked R are equal, and
the output is an inverted copy of the input, with the same amplitude.

selective fading a form of FADING in which some frequencies of a signal are affected more than others, causing severe distortion that cannot be remedied by devices such as AUTOMATIC GAIN CONTROL.

selective interference a form of INTERFERENCE such as deliber-ate jamming, that affects one or a few frequencies.

selectivity the ability to select a particular FREQUENCY. The sim-plest measurement of selectivity is the BANDWIDTH of the circuit or complete device.

See also Q-FACTOR.

selenium an element related to sulfur that was one of the first known SEMICONDUCTORS. It exhibits PHOTOCONDUCTIVITY and when in contact with a metal acts as a PHOTOVOLTAIC CELL.

selenium rectifier a RECTIFIER that uses selenium-iron junctions. Selenium rectifiers once were used extensively for power supplies. They have now been replaced with the smaller and more efficient

SILICON DIODES. An incidental advantage of this is the elimination of the hazard of the poisonous fumes given off by burning selenium that used to be a predominant smell in electronics assembly facilities.

self-capacitance the STRAY CAPACITANCE between adjacent turns of wire in a coil. This causes a coil to have a self-resonant frequency, and an equivalent single value of capacitance can be calculated from the inductance value and the resonant frequency.

self-excited (of an oscillator) self-starting. Some very stable oscillators are not self-starting and need an applied signal to start oscillation.

self-inductance the quantity that relates the MAGNETIC FLUX of an inductor to the current flowing in the windings of that inductor. Symbol: L; unit: henry. For a given size and shape of conductor in the absence of FERROMAGNETIC material, the inductance value is constant.

self-sealing capacitor a CAPACITOR that resists damage by excessive voltage. If the DIELECTRIC is punctured, localized heating oxidizes the metal electrode around the puncture, restoring the insulation.

SEM *abbreviation for* scanning electron microscope. See ELECTRON MICROSCOPE.

semiconductor a material whose CONDUCTIVITY can be controlled by the presence of IMPURITIES. The conductivity of a doped semiconductor (see DOPING) may be one or more orders of magnitude lower than that of a conductor. Semiconductors can be elements, such as selenium, germanium, or silicon, or compounds such as gallium arsenide, antimony trisulfide, or cadmium sulfide. In the pure state, semiconductors are reasonably good insulators, but they have a NEGATIVE TEMPERATURE COEFFICIENT of RESISTIVITY unlike metals.

Traces of impurity have a considerable effect on resistivity, lowering the value to levels that can approach the resistivity values of some metals. In addition, the type of MAJORITY CARRIER whether electron or hole, can be determined by the type of material used as an impurity, or doping material. See INTRINSIC, DONOR, ACCEPTOR, DIFFUSION, JUNCTION (2).

semiconductor diode a DIODE that makes use of a SEMICONDUC-

TOR material. JUNCTION (2) diodes make use of the contact between n-type and p-type semiconductors. POINT-CONTACT DIODES use a metal point contact to a semiconductor surface. SCHOTTKY DIODES use a metal in contact with a larger area of semiconductor.

semitransparent cathode a PHOTOCATHODE partly transparent to the light that causes EMISSION. Such a material can be used in the form of a thin film on a glass plate, with light striking the surface from the glass side, and electron emission from the other side. Such photocathodes were used in early types of TV CAMERA TUBES and are still used in PHOTOMULTIPLIERS.

sense wire the wire that threads a MAGNETIC CORE and detects a change of magnetization in the core-storage system of computer memory.

sensitivity the ratio of OUTPUT change to INPUT (1) change for a measuring device. For a measuring instrument, sensitivity is given in terms of the amount of input for FULL-SCALE DEFLECTION or the amount of input for unit deflection. For example, a milliammeter may be designated as 1.0 mA for fsd (full-scale deflection), and an oscilloscope amplifier as 1 V per cm. For a radio receiver, the sensitivity would be given as the minimum input signal in microvolts needed to ensure a useful signal-to-noise ratio.

sensor or **input transducer** any type of measuring or detecting element or TRANSDUCER. A sensor will detect changes in the quantity to which it is sensitive, for example, light, temperature, strain, or rotation, and transmit this change as an electrical signal.

sequential control the control of a device by a sequence of instructions. MICROPROCESSOR-controlled equipment uses sequential control, and the controlling sequence of instructions is known as the program.

sequential field an early type of COLOR TELEVISION system. In a sequential field system, a complete picture in each primary light color was transmitted and shown in sequence. The rate of display had to be fast enough to appear as one picture to the eye. Field-sequential methods, in which three complete fields are used per picture, are not compatible with MONOCHROME reception, but are still used for narrow-bandwidth color TV. See also LINE-SEQUENTIAL TELEVISION.

sequential logic a LOGIC system in which the output depends on the sequence in which inputs occur rather than on the instantaneous combination of inputs. The FLIP-FLOP is a sequential device, because its output, unlike that of a GATE is usually determined by the sequence of past inputs rather than a present input. For example, an R-S FLIP-FLOP with both inputs at zero can have an output that is 1 or 0, depending on what the *previous* input was. Contrast this with a gate, in which the output is determined completely by the current inputs.

serial counter see ASYNCHRONOUS COUNTER.

serial memory a MEMORY that employs devices such as FLIP-FLOPS connected in series. A set of flip-flops in series, a SERIAL REGISTER, can be clocked by a common PULSE so that an input at one end can be shifted through the register by clocking. The data bit moves along by one flip-flop for each clock pulse. Data stored in such a register can be obtained only by clocking out all the previously stored data. Modern computers make more use of RANDOM-ACCESS MEMORY systems.

serial transfer the movement of digital data in binary form along a single cable. Each bit is transferred as a logic 0 or 1 signal along the cable. For a group of bits, the transfer is controlled by a master clock rate. See also SERIAL TRANSMISSION.

serial transmission the transmission of digital data along a single serial link. The transmission can be SYNCHRONOUS or ASYNCHRONOUS and both types rely on generating identical clock pulse rates. In the synchronous system, the sending terminal sends out a synchronizing character, a set of 1s and 0s. This is used by the receiver to achieve synchronization and from that time on, the two are held in synchronization by transmitting either data or the synchronizing character. In the asynchronous system, data are sent at intervals, with nothing on the line when there are no data. Each group of data, usually each byte, carries its own synchronizing signals in the form of a start bit and two stop bits.

series the connection of COMPONENTS so that the same current flows through each in turn. See also PARALLEL.

series network a NETWORK in which components are connected in series with the signal.

series-parallel (of a CIRCUIT or NETWORK) having some components in series with the signal and others in parallel.

series resonance a RESONANT CIRCUIT in which the impedance becomes a minimum at resonance. The series connection of an inductor and a capacitor is the simplest series resonant circuit. At resonance, the impedance of this circuit is entirely resistive, and voltage is in phase with current.

series stabilization a method of stabilizing CURRENT or VOLTAGE that uses a controller, such as a transistor, in series.

service area the area within which a TRANSMITTER can be usefully received. The area is shown as a set of field-strength contours on a map and is often divided into primary and secondary regions. Outside the secondary region, reception is unsatisfactory either because of severe fading or interference from other transmitters.

servo any mechanical system that is electrically controlled, particularly a motor that is remotely controlled and will duplicate the movements of a master generator shaft.

seven-segment display a method of displaying digits and some letters with seven bar shapes. The seven-segment display is extensively used for LIGHT-EMITTING DIODE and LIQUID-CRYSTAL DISPLAYS particularly for calculators. The alternative is a DOT-MATRIX type of display. See Fig. 84.

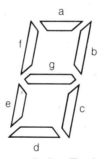

FIG. 84. **Seven-segment display.** The form of a seven-segment display, showing the conventional lettering system for the segments. Some displays include decimal points in addition to the segments.

shaded-pole motor a type of AC motor in which a phase shift is created in the field by a copper ring placed over part of the field core. Presence of this PHASE SHIFT creates a rotating field, and the metal rotor will try to keep in step with the rotation at the frequency of the AC supply.

shading an unwanted AMPLITUDE change in a TV SIGNAL. Shading usually causes a uniformly lit scene to appear to contain shadows when the image is viewed.

shadow the ATTENUATION of radio signals caused by obstructions between the TRANSMITTER and the RECEIVER. Hills, power lines, and large metal objects can all cause shadowing, and the effect can be severe at the higher frequencies.

For TV signals, improvements can sometimes be achieved by using a DIRECTIONAL ANTENNA pointed to a local reflection, but in very bad cases it may be necessary to use a REPEATER of some sort. A passive repeater, consisting of a receiving antenna on one side of a hill connected to a transmitting antenna on the other, is often successful. Direct broadcasting from satellites (*DBS*) eliminates problems of shadowing.

shadow mask the original type of TV receiver color tube. The shadow mask was made of nickel-steel, with a low expansivity, and was perforated with holes in groups of three, using a triangular arrangement. The holes, combined with the deposition of phosphor dots on the screen in the same pattern, ensured that electrons from three separate guns arrived at the correct PHOSPHOR dots. The guns were also arranged in a triangular structure.

The weakness of the system lay in the number of controls and adjustments that were needed in order to preserve good CONVERGENCE since each control had to deal with both vertical and horizontal effects. Very often, convergence had to be reestablished if a receiver was moved. The later development of the APERTURE MASK greatly reduced convergence problems, because it allowed the three guns to be constructed in line, thus making all corrections along one axis only.

Shannon's sampling theorem a theorem stating that an analog signal must be sampled at least twice as fast as the highest frequency component contained within the signal if ALIASING errors are to be avoided.

shaped-beam tube a form of display cathode-ray tube for alphanumeric displays. The characters are formed by acting directly on the shape of the beam with electric or magnetic fields, rather than by scanning of the beam.

SHF *abbreviation for* superhigh frequency. This refers to the range 3 to 30 GHz (wavelength 1 to 10 cm) in the MICROWAVE range.

shifting the action of moving binary digits in a REGISTER. In a shift action, the bits are moved from one flip-flop to the next in line. The bit at one end of the register will be lost, and a zero bit will be loaded in at the other end.

shift register a set of FLIP-FLOPS connected in series. The flip-flops are connected with the output of one taken as the input to the next. When a clock pulse is applied to all of the flip-flops together, the bit (0 or 1) in each flip-flop will be transferred to the next in line. The connections can be made either for left-shift or for right-shift, or by way of gates so that both left- and right-shift can be obtained by using a shift-direction signal. The shift register is the basis of SERIAL MEMORY.

shock excitation the production of oscillations by a pulse. This causes RINGING (1).

Shockley diode a general name for a pnpn diode, such as a THY-RISTOR or SCS.

short see SHORT CIRCUIT.

short circuit a low-RESISTANCE connection, both wanted and unwanted. Sometimes abbreviated to *short*. The effect of a short circuit is to equalize voltages at two points and to allow current to flow. If the short circuit is a fault condition, it will usually cause problems such as loss of signal, loss of control, or damage to components. Deliberate short circuiting is used in signal switching, or for making high-voltage circuits safe to touch.

short-circuit impedance a measurement on a NETWORK. The short-circuit IMPEDANCE at the input of a network is the impedance at the input when the output is short-circuited to signals. The short-circuit impedance at the output is the output impedance when the input is short-circuited to signals.

short wave a radio signal in the frequency range of about 3 to 30 MHz. This range contains many of the most popular amateur bands, as well as many of the international communications channels.

short-wave converter an add-on circuit to enable SHORT-WAVE reception by a MEDIUM-WAVE receiver. The converter contains a local oscillator and mixer that convert short-wave signals into signals in the medium-wave band. The receiver thus becomes the second part of a DOUBLE-SUPERHETERODYNE RECEIVER.

shot noise see SCHOTTKY NOISE.

shunt 1. a COMPONENT connected in PARALLEL with a SIGNAL. A *meter shunt* is a low-value RESISTOR connected in parallel with the movement in order to allow measurement of currents greater than thefull-scale deflection current of the movement. **2.** *vb.* connect a COMPONENT in PARALLEL with another.

shunt-derived filter a form of FILTER in which parallel (SHUNT) resonant circuits are used.

shunt feedback or **voltage feedback** the FEEDBACK of a part of the VOLTAGE output to the input of an amplifier.

shunt network a NETWORK connected in PARALLEL with a SIGNAL.

shunt stabilization the STABILIZATION usually of voltage, by a device connected in parallel with the load. The use of a ZENER DIODE in parallel with a voltage source is an example of shunt voltage stabilization. The stable voltage is the voltage across the diode, with a series resistor used as a load. Variations in the supply voltage cause current variations that appear as voltage changes across the resistor, not across the zener diode. Changes in the load current cause changes in the current through the zener diode, with negligible effect on the voltage output.

The stabilization breaks down if the current through the zener diode falls below the stabilizing limit, or if the supply voltage is reduced to the level of the stabilized voltage. See Fig. 85.

sideband a range of frequencies caused by the MODULATION of a carrier. Any form of modulation of a carrier will result in the appearance of sidebands. One set, the upper sideband, will be at higher than carrier frequency, with the other set, the lower sideband, at a symmetrically lower frequency. For AMPLITUDE MODULATION the sideband frequencies are directly related to the modulating frequency. Each upper-sideband frequency is the sum of the carrier frequency and a modulating frequency, and each lower-sideband signal is the difference between the carrier fre-

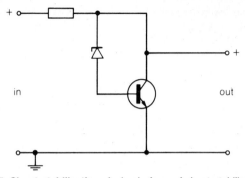

FIG. 85. **Shunt stabilization.** A simple form of shunt stabilizer, another form of the AMPLIFIED-ZENER circuit. The total current is divided between the transistor and the load (not shown), and the voltage is maintained constant by the action of the zener diode. If the load current increases, the transistor current decreases, keeping the total current constant.

quency and a modulating frequency. The useful part of the signal is carried in the sidebands, but these together constitute only up to 50% of the total signal energy. For frequency modulation, the inner sidebands have the same frequencies as for AM, but there are also outer sidebands unless the MODULATION INDEX is small.

side frequency a single frequency in a SIDEBAND.

side lobe a region of sensitivity or signal strength to the side of the required directional axis of an ANTENNA. Most DIRECTIONAL ANTENNAS allow some side-lobe formation, but this is harmless unless it causes interference.

siemens or (formerly) **mho** the unit of CONDUCTANCE that is equal to the inverse ohm. Symbol: S.

signal a variation in VOLTAGE or CURRENT that is used for carrying information.

signal electrode an ELECTRODE from which a signal output is taken. The term usually denotes the signal output electrodes of TV CAMERA TUBES, PHOTOMULTIPLIERS, and TRANSDUCERS. See BACKPLATE.

signal generator a laboratory instrument that is a combination of controllable calibrated oscillator (see CALIBRATION) and AT-

TENUATOR. Signal generators are used to supply signals of known amplitude at known frequencies for test purposes. See also PULSE GENERATOR.

signal level the AMPLITUDE of a SIGNAL. This is often given relative to some standard, such as the 1V video signal. When the standard level is well known, the signal level is often given in (voltage) dB.

signal-to-noise ratio (S/N) the ratio of levels of desired SIGNAL to unwanted NOISE in decibels. The usefulness of a received signal is determined by its S/N. An acceptable video signal needs an S/N in the region of 45 dB, and a hi-fi audio signal requires at least 60 dB.

signal winding the WINDING of a MAGNETIC AMPLIFIER to which the low-frequency AC signal is applied.

silent zone the area around a TRANSMITTER that receives neither the ground wave nor a strong echo. See also SKIP DISTANCE.

silicon a semiconducting ELEMENT obtained from sand.

silicon cell a PHOTOCELL that uses silicon as its sensitive material.

silicon chip a small thin rectangular piece of silicon, typically 2 mm $\times$ 1 mm $\times$ 0.1 mm, typicall cut from a thin circular WAFER.

silicon-controlled rectifier see THYRISTOR.

silicon-controlled switch (SCS) a four-layer (pnpn) DIODE in which connections are made to the inner P-LAYERS and N-LAYERS. A positive voltage is applied to the anode, which is the outer p-layer. The cathode, the outer n-layer, is at zero voltage. In this condition, a pulse to the first gate, the inner p-layer, will switch the device on. A positive pulse on the second gate, the inner n-layer, will switch current off.

The SCS is used for low-power switching as compared with the THYRISTOR which has a similar structure and can be used only for comparatively low switching rates because of its comparatively long turn-on and turn-off times.

silicon diode a DIODE that uses a rectifying silicon junction. The FORWARD DROP is rather high, of the order of 0.6 V, but the resistance in the REVERSE DIRECTION is also high. Silicon diodes are used extensively in manufacturing because of their low price.

silicone grease a synthetic grease that is an excellent electrical INSULATOR and a reasonably good heat CONDUCTOR. The grease is similar in chemical structure to a natural grease but with silicon substituted for carbon, hence the name.

silicon-on-sapphire (SOS) a construction technique for metal-oxide semiconductor INTEGRATED CIRCUITS used particularly for military and other applications where high reliability is required. Synthetic sapphire is used as a SUBSTRATE and silicon is deposited on it by EPITAXY. The resulting chips have a high performance, but the construction is difficult and costly.

silicon rectifier a SILICON DIODE of any type used as a RECTIFIER.

silver mica the highest quality of mica used for mica capacitors.

simplex a data channel that allows communication in one direction only. Compare DUPLEX, HALF-DUPLEX.

simulator a SYSTEM usually computer-based, that imitates the behavior of another system. For example, the simulator can imitate the reaction of a NETWORK to a PULSE the response of an aircraft to wind shear, the reaction of an office organization to the death of a manager.

The importance of simulation is that any reasonably well understood system can be simulated, and the simulation used to predict behavior of the main system. For a partly understood system, the simulation may be used to gain understanding by comparing the predictions of the simulator with the actual outcome of an experiment.

sine/cosine potentiometer a POTENTIOMETER seldom encountered in modern use, in which the DIVISION RATIO is proportional to the sine or cosine of the angle of rotation of the shaft.

sine wave a WAVEFORM whose shape is that of a graph of the sine of an angle plotted against the angle. A sine wave is the natural waveform of voltage generated by a coil of wire revolving in a uniform magnetic field.

singing an OSCILLATION with an audible OUTPUT.

single-ended or **unbalanced** (of a circuit) having one terminal at ground at both input and output. The signal at the input or output consists of a voltage that will vary with respect to ground. In a BALANCED AMPLIFIER the sum of the signals at two inputs or outputs is always zero.

single-phase (of an AC system) having one supply conductor and a return.

single-pole switch a switch that changes the connection of one line only.

single-shot see MONOSTABLE.

single-sideband (SSB) (of a transmitter or receiver) transmitting or receiving radio signals with one SIDEBAND only. In a conventional amplitude-modulated signal, all information in the modulating signal is held in each one of the SIDEBANDS. This means there is redundancy of signal that can be reduced by transmitting only one sideband. Since one sideband represents only a quarter of the total transmitter power of a double sideband system, this implies that the transmitted power can be reduced with no loss in reception range. By eliminating the carrier as well, all the power of the transmitter can be devoted to the information-bearing sideband. The drawback is that much more elaborate transmitters and receivers are needed.

single-throw switch see DOUBLE-THROW SWITCH.

sink 1. accept CURRENT from a TERMINAL. TRANSISTOR-TRANSISTOR LOGIC integrated circuits require any driving device to be able to accept (sink) a current of up to 1.6 mA to ground when a terminal is at logic 0. 2. *n*. see HEAT SINK.

sinusoidal (of a WAVEFORM) having the shape of a SINE WAVE.

SIPO register a serial-in, parallel-out REGISTER in which data are fed in serially, using one clock pulse per bit of data, and then read out on a set of parallel lines, using one line output for each FLIP-FLOP in the register. See Fig. 86. Compare SISO REGISTER, PIPO REGISTER, PISO REGISTER.

SISO register a serial-in, serial-out REGISTER in which bits are clocked in, travel through the register, and are clocked out of the other terminal. See Fig. 87. Compare SIPO REGISTER.

SI unit any of the units adopted for international use (Système Internationale) in science and engineering, based on the meter, kilogram, second, and ampere.

skeleton slot a type of SLOT ANTENNA in which the metal surrounding the slot has been reduced to a tube or wire frame.

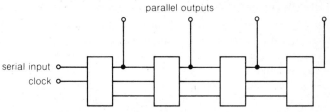

FIG. 86. **SIPO register.** The set and reset terminals have been omitted for clarity.

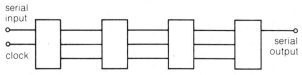

FIG. 87. **SISO register.** The set and reset terminals have been omitted for clarity.

skew the time difference between signals, particularly PULSE signals.

skin effect the conduction of high-frequency currents on the outside surface of a CONDUCTOR. Conductors for high-frequency currents can be made of thick wire or tube, and use silver plating to increase conductivity.

skip distance the distance between a transmitter and the region in which the first reflection from the IONOSPHERE is received. There will be a SILENT ZONE in the area that lies between the skip distance and the greatest distance covered by the ground wave.

sky wave or **reflected wave** the DIRECT WAVE that reaches a RECEIVER from a TRANSMITTER after reflection from the IONOSPHERE. Compare GROUND WAVE.

slave circuit a LOGIC CIRCUIT that is solely driven by the output of another and reproduces the output waveform. The slave portion of a MASTER-SLAVE FLIP-FLOP for example, reproduces a voltage level from the master section but is clocked at a different time.

slave processor or **coprocessor** a MICROPROCESSOR in a computer or controller circuit that is totally controlled by a main microprocessor. The usual type of slave processor is used for memory access, screen control, or mathematical operations.

slew rate the rate of change of SIGNAL voltage. This is a better guide to the frequency response of transistors and amplifiers, particularly OPERATIONAL AMPLIFIERS, to large signals than the use of high-frequency cutoff figures, which only relate to small signals.

Slew rate is important in feedback amplifiers because of the possibility of breakdown of feedback. If the slew rate is slower than the response of the feedback loop, for example, the amplifier will effectively be operating without feedback during the time of rapid voltage change at the input.

slope resistance see INCREMENTAL RESISTANCE.

slot antenna an ANTENNA that consists of a rectangular slot in a metal sheet or in a metal mesh. The long dimension of the slot is along the plane of the magnetic field, so the transmitted wave is POLARIZED (1) at 90 degrees to the polarization of a conventional DIPOLE ANTENNA. The balanced feed is to or from the midpoints of the edge of the slot, and the CHARACTERISTIC IMPEDANCE is approximately 500 ohms. See also SKELETON SLOT.

slotted waveguide a WAVEGUIDE into which slots are cut in places to allow a PROBE (1) to be inserted.

slow-break switch a SWITCH that does not use spring-assisted snap-over action. An example is the old-fashioned KNIFE SWITCH.

slow wave an electromagnetic wave that has been slowed to about the speed of electrons in a beam. The slowing can be achieved by propagating the wave in a spiral path, as in a TRAVELING-WAVE TUBE.

slugged relay a form of RELAY in which a copper loop, the slug, has been placed over the magnetic circuit. This makes the action of the relay slower because of the effect of EDDY CURRENTS.

slug tuning see PERMEABILITY TUNING.

small-signal gain the GAIN figure of a system obtained from the ratio of output to input signal amplitudes for signals that are of small amplitude relative to the power supply voltage.

small-signal parameters the PARAMETERS such as input resistance and transfer current ratio that are measured for small-signal, not DC, quantities. The term denotes particularly the parameters used to specify transistor action.

smoothing 1. the removal of unwanted AC content from DC. 2. the filtering of any signal by a LOW-PASS FILTER.

SMPTE *abbreviation for* Society of Motion Picture and Television Engineers.

snap-off diode see CHARGE-STORAGE DIODE.

snow a slang term for the effect of NOISE on a TV or radar picture signal.

socket a female connector. Compare PLUG.

soft magnetic material a material with high PERMEABILITY but low RETENTIVITY. Such a material will become strongly magnetized when in the magnetic field of a coil or other magnet but retains practically no magnetism when the field is removed. See MOVING-IRON METER.

software the programs for a MICROPROCESSOR-controlled system. Compare HARDWARE.

soft x-ray an X-RAY beam that has been created using a comparatively low-energy electron beam. Soft x-rays have a lower frequency range than the more penetrating hard x-rays.

solar cell a PHOTOVOLTAIC CELL used to obtain electrical power from light. Cells based on gallium arsenide have been used for satellites and spacecraft to provide power over long periods. The voltage output is low, a fraction of a volt per cell, and the efficiency is also poor, about 10%. When no practical alternative exists, however, solar cells can provide useful power.

solar noise the radio NOISE generated from the random movement of ions in the sun.

soldering the joining of metals with a low-melting alloy of tin and zinc known as solder. Soldering is the most common method of making mechanical and electrical connections to electronic components. For hand work and repair work, an electrically heated soldering iron is used, along with a solder alloy that is in the form of hollow wire, with a core of flux. The flux is a resin that spreads over the metal being soldered to prevent oxidation. Mass-produced printed circuit boards are soldered using a HOT-DIP system. For highly reliable systems, assemblies are made by robot methods, using spot-welding rather than soldering. See also DIP-SOL-DERING.

solenoid 1. a tubular coil of wire whose length is large compared with its diameter. 2. an electromechanical actuator that consists of a coil with a movable core. When current flows in the coil, the core is drawn into the coil, enabling use of the mechanical action.

solid conductor a single-core wire, as opposed to stranded or hollow-cored wire.

solid-state circuit any CIRCUIT employing semiconductors as active components, with no mechanical moving parts or any vacuum devices other than a CATHODE-RAY TUBE.

solid-state camera a TELEVISION CAMERA that dispenses with the conventional vacuum tubes (VIDICONS) for image pickup. Of the several methods tried, the most promising and practical has been the use of CHARGE-COUPLED DEVICES as photosensors. It allows charge that has accumulated in a row of CCD cells to be read out in the time of a line flyback pulse by a number of clock pulses. A few simple CCD cameras have appeared for use with computers, but their major role to date has been in TV systems for use in space, and as a result of this work, such cameras can now be obtained as part of a video recorder system.

solid-state memory a MEMORY system that uses semiconductors as distinct from MAGNETIC CORE (2) devices. See also ROM, RAM, DYNAMIC RAM.

solid-state physics the branch of physics that deals with the structure and behavior of solids, particularly SEMICONDUCTORS.

sonar a form of acoustic RADAR. ULTRASONIC waves in water are used to locate objects in the water, employing the reflection from a pulse output. The principles are used in echo sounding.

sound any audible pressure waves in air. Sound waves can also be transmitted at higher speeds in liquids and in solids. The frequencies above the range of hearing are called ULTRASONIC. Those below the range of hearing are called SUBSONIC signals.

sound carrier the radio frequency CARRIER (1) for the sound of a TV picture. In the PAL system, the sound is frequency modulated on to a separate carrier whose frequency is 6 MHz higher than the vision carrier. See also INTERCARRIER SOUND.

sound recording see RECORDING OF SOUND.

sound reproduction the recovery of audio signals from a recording and their conversion into sound. A sound reproduction system consists of a TRANSDUCER amplifier, and loudspeaker system.

A general description of a good hi-fi system shows what is required for each type of reproduction system. The transducer converts the recorded sound into audio signals.

For the most common form of recording, the mechanical disk, the transducer is a PICKUP head. It uses a stylus fitting into the groove on the disk, which vibrates, driven by the modulation of the groove. The disk has to be spun at a steady speed, generally 33 and 1/3 revolutions per minute, though 45 revolutions per minute is used for shorter recordings. For a stereo recording, this modulation will be picked up as vibration in two mutually perpendicular directions at 90 degrees to each other. Transducers connected to the stylus will then convert these vibrations into electrical signals. The transducers can use PIEZOELECTRIC CRYSTALS or the MOVING-IRON PICKUP or MOVING-COIL PICKUP principles. The small signals from the pickup are voltage-amplified by a PREAMP and any corrections, such as EQUALIZATION, are performed.

If the source of signals is tape, then a tape transcriptor must be used to pull the tape at a constant and correct speed past the replay head. The fluctuations of magnetic field on the tape are converted into induced voltages in the head, and these signals are deemphasized as required in the preamplifier. The signals from a cassette unit may have to be expanded (see COMPANDER) if they have been compressed during recording for noise-reduction purposes.

For COMPACT DISKS the speed of the disk is controlled by a motor in such a way as to present a constant linear speed to the pickup unit. The start of the recording is in the inner track, and reading is carried out by a solid-state LASER and optical system. The reflection of light from a flat surface indicates that a 0 is present, its scattering by an etched pit indicates a logic 1. The digital bit signals are detected by a photocell, and these signals are amplified before being passed to the preamp unit.

For radio signals, a TUNER is used. This will normally be an FM tuner, though several designs allow for medium wave reception also. The carrier from the antenna is tuned, converted to IF, amplified, and demodulated. The audio signal is deemphasized (see DEEMPHASIS), and then passed to the preamp.

At the preamp, a switch enables the user to select any of the sound sources for reproduction. Simple systems make use of mechanical switches, but quality systems use the switch to operate separate MOSFET switches (see FIELD-EFFECT TRANSISTOR), so switch pulse noise and switch contact noise are eliminated. The signal is then voltage-amplified, and potentiometer controls for gain, bass, and treble are included. The signal then passes to the main amplifier in which several stages of voltage amplification are followed by a power amplifier stage. This provides an output with enough power to drive a loudspeaker, usually of the magnetic type. See also SOUNDTRACK, RECORDING OF SOUND.

soundtrack the edge of a motion picture film that is used to carry the sound MODULATION. The sound is modulated in the form of variable density bands in which density is proportional to amplitude. Because the film passes through the projection gate in PULSES being held for about 1/24 second, and then rapidly jerked on, the sound recording and reproduction must be done at a different part of the film, which is moving steadily. This part of the film is ahead of the picture, following the loop that enables the film to regain steady motion after its intermittent movement through the projection gate. Because of this, any splicing because of film breakage can cause sound synchronization to be lost.

The SIGNAL-TO-NOISE RATIO for film sound is poor unless methods such as the DOLBY SYSTEM are used. The reproduction of film sound makes use of a light source and a PHOTOCELL. The photocell must be capable of reasonably fast response, so sensitivity has to be sacrificed and a high-gain amplifier must be used for the output signals. Magnetic tape is also used for film soundtracks.

source 1. (a) an emitter of CHARGE CARRIERS in a FIELD-EFFECT TRANSISTOR. (b) any device that produces signals. 2. *vb.* provide CURRENT to drive a LOAD for example, drivers for logic devices. See also SINK (1).

source follower a FIELD-EFFECT TRANSISTOR common-drain amplifier circuit. This is the FET equivalent of an EMITTER FOLLOWER with high input impedance and low output impedance.

source impedance the complex ratio of voltage amplitude and phase to current amplitude and phase for any source of signal (see IMPEDANCE).

space charge a cloud of ELECTRONS or other CHARGE CARRIERS originally denoting the electrons around the hot CATHODE of a vacuum emitter. Space charges also exist in semiconductors around a junction. Because the charge acts as a POTENTIAL BARRIER to the movement of electrons, its existence causes a form of bias to be present even when no external voltage is applied.

space-charge density a measure of the SPACE-CHARGE effect equal to the amount of charge per unit volume in a space-charge region.

space-charge limiting the limiting of CURRENT because of the effect of a SPACE CHARGE. This type of limiting means that the electron current drawn from the region around the cathode is not greatly affected by variations in the cathode temperature, provided that the cathode can maintain the space charge. Most transmitting tubes operate under space-charge limited conditions rather than in the alternative temperature limiting conditions.

spark an electrical discharge caused by the breakdown of INSULATION in air. Sparking in air can occur more easily if the air is ionized (see IONIZATION), for example, by radioactive particles.

spark counter a detector for alpha particles. Alpha particles have high ionizing (see IONIZATION) capability but are stopped by thin layers of materials of ordinary density. A spark counter detects alpha particles in air by using a high voltage between an ANODE and a CATHODE. The presence of an alpha particle will ionize the air and cause a spark. If the anode is supplied through a large value load resistor, there will be a voltage pulse at the anode that can be amplified and used as the input to a counter.

spark gap an arrangement of ELECTRODES to encourage a SPARK to pass under fault conditions. The arrangement is used to protect devices such as CATHODE-RAY TUBES from voltage surges that would otherwise cause internal sparking. See SURGE DIVERTER.

spark-suppression circuit a CIRCUIT intended to reduce SPARKS at switch or relay contacts. Typical circuits use capacitor-resistor networks that form a damped oscillating circuit, reducing the rate of change of voltage as the contacts open. Spark suppression is particularly important if the contacts switch an inductive load, and suppression using diodes is then more common.

speaker see LOUDSPEAKER.

spectrum a range of FREQUENCIES or WAVELENGTHS. The electromagnetic spectrum consists of all electromagnetic waves, from the very low frequencies up to the gamma-ray range and beyond. The term is often used to denote a band of frequencies, for example, *microwave spectrum.*

spectrum analyzer an instrument whose output is a graph that shows a plot of amplitude or phase against frequency for a complete FREQUENCY BAND.

speech bandwidth the minimum BANDWIDTH needed for intelligible speech, about 300 Hz to 2.7 kHz.

speech recognition a method of controlling a computer by spoken commands. The voice sounds are picked up by a MICROPHONE and processed into digital codes that are compared with standardized stored codes. If an input causes a code to be generated that is sufficiently similar to a stored code, the action corresponding to the stored code will be executed. See also RECORDING OF SOUND.

spider formerly, a metal or plastic flat spring that located a LOUD-SPEAKER cone and SPEECH COIL in place. The term arose from the traditional shape of the spring.

spiral PDA a form of POST-DEFLECTION ACCELERATION that uses a spiral of highly resistive film.

spiral scanning a form of CATHODE-RAY TUBE scanning. A spiral SCAN starts at the center of the face of the tube and traces out a spiral path until it reaches the edges.

sporadic-E effect an effect that causes unusual PROPAGATION conditions. At times, highly ionized (see IONIZATION) layers form in the E LAYER about 100 km above Earth. This heavily ionized layer reflects frequencies that normally pass through the IONO-SPHERE and causes long-distance reception of signals, such as VHF TV signals whose range is normally line of sight. This in turn causes severe INTERFERENCE between transmissions that are normally not within range of each other. Sporadic-E conditions are associated with intense sun-spot conditions and are usually predictable. See also SUNSPOT CYCLE.

spot 1. the area on the screen of a CATHODE-RAY TUBE that is lit by the electron beam. 2. an imperfection on a screen that will not illuminate.

spot speed the rate of scanning of a CATHODE-RAY TUBE in terms of SPOT (1) diameters per second.

spreading resistance the internal RESISTANCE of collector or base connection attributed to the semiconductor material between the junction and the contact.

spurious response an unwanted signal, particularly in a RADAR system.

sputtering a coating method employing a low-pressure gas discharge in which the material to be coated is the ANODE, and the coating material is the CATHODE. Any material can be coated by sputtering, and almost any refractory coating material can be used, for example, tungsten, platinum, diamond (carbon), and silica.

square-law detector a type of amplitude DEMODULATOR. The output voltage of a square-law detector is proportional to the square of the input voltage. A small amount of carrier modulation can therefore produce a comparatively large output from the circuit, so that the detector is sensitive. Metal-oxide semiconductor devices are particularly good square-law detectors.

square-loop hysteresis a characteristic shape of HYSTERESIS LOOP that is straight-sided.

square wave a WAVEFORM with a square shape. The waveform has vertical sides, representing a rapid switch between low voltage and high voltage. The time spent at high voltage is equal to the time spent at low voltage, making the MARK-SPACE RATIO equal to unity. Waves with steep sides that do not have unity mark-space ratio should be described as *rectangular waves* but are generally known as square waves unless the mark-space ratio is very low, when the waves are called PULSES.

square-wave response the OUTPUT of a CIRCUIT for a SQUARE-WAVE input. Poor high-frequency response, low SLEW RATE, shows up as sloping sides in the output wave, and poor low-frequency response as a sagging top (see SAG). Any instability also shows itself as RINGING (1) after a change of voltage.

squegging (of an OSCILLATOR) biasing off at intervals. A squegging oscillator uses too much FEEDBACK driving the circuit into NON-LINEAR conditions and causing a capacitor in the circuit to charge, thus biasing off the oscillator. The oscillation then resumes when the capacitor has discharged. The BLOCKING OSCILLATOR is an intentionally designed type of squegging oscillator.

squelch a CIRCUIT technique for reducing interstation NOISE. The voltage on the AUTOMATIC GAIN CONTROL line is used to suppress the audio stages unless a signal is tuned that is strong enough to operate the line. This enables a radio to be tuned from one station to another without hearing the noise between carriers. The squelch technique has obvious benefits, for example, in car radio reception.

SSB see SINGLE SIDEBAND.

SSI *abbreviation for* small-scale integration, that is, about 10 active devices per chip.

stabilization a CIRCUIT technique for making a quantity resist changes in INPUTS or LOADS. VOLTAGE STABILIZERS are circuits whose output voltage remains stable despite changes in load current or supply voltage. *Current stabilizers* are circuits that supply a constant current despite changes in load resistance or supply voltage. Stabilization of the GAIN of an amplifier against changes in component values can be achieved by using NEGATIVE FEEDBACK. A circuit can also be said to be stabilized if phase correction is used to prevent unwanted oscillations or RINGING (1).

stable circuit a circuit that will not break into oscillation under normal working conditions. If the circuit is truly stable, it will not show RINGING (1) under SQUARE WAVE testing. Compare BISTABLE CIRCUIT.

stage a clearly defined part of an AMPLIFIER such as the circuitry associated with a transistor or integrated circuit.

stage efficiency the ratio of power output from a stage to the power that has to be supplied to it (DC supply as well as signal power), used mainly of transmitter stages.

stage gain the gain, usually VOLTAGE GAIN of a STAGE. This is often expressed in decibels, using $20 \log (V_{out} / V_{in})$.

stagger tuning a method of obtaining a large BANDWIDTH in a TUNED AMPLIFIER. Several tuned circuits in different stages of an amplifier are tuned to slightly different frequencies, so the overall tuning is wideband.

staircase voltage a voltage waveform that takes the form of a RAMP with a number of level sections so that the shape is that of a flight of stairs. See Fig. 88. See also DIGITAL VOLTMETER.

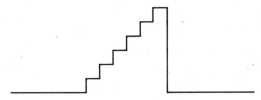

FIG. 88. **Staircase voltage.** The appearance of a typical staircase waveform.

standardization the setting up of a basis of comparison of units of measurement. Standardization of an instrument, such as a voltmeter or oscilloscope, means that its readings are compared with those of a standard, usually a SECONDARY STANDARD. The instrument is then corrected or calibrated. The term is also used to denote the establishment of written standards for the manufacturing, testing, and specification of components and devices.

standing wave or **stationary wave** a wave fixed in space, so the amplitude of wave oscillation at any given point along the wave's axis is fixed for as long as the wave pattern persists.

A NODE of the standing-wave pattern is a point of minimum amplitude, and an ANTINODE is a point of maximum amplitude. The pattern has the same wavelength as the wave that generates it, so wavelength can be measured from the pattern (see LECHER WIRES). Compare TRAVELING WAVE.

standing-wave ratio (SWR) a measurement of the efficiency of line transmission that applies also to WAVEGUIDES and ANTENNAS.

STAR-DELTA TRANSFORMATION

The SWR is the ratio of the maximum amplitude at any point on the line to the minimum amplitude. The smaller the SWR—the minimum possible value is unity—the better the MATCHING of the line. See also SWR METER.

star-delta transformation a theorem for finding the equivalent component values for star and delta arrangements of connections for THREE-PHASE SUPPLY AC. The principle states that the star network (see Fig. 89a) is equivalent to the delta network (see Fig. 89b), with component values given by the following equations:

$$y_1 = Y_1 Y_3 / S$$

$$y_2 = Y_2 Y_3 / S \qquad \text{where } S \text{ (the sum)} = Y_1 + Y_2 + Y_3$$

$$y_3 = Y_1 Y_2 / S$$

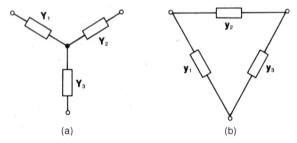

(a) (b)

FIG. 89. **Star-delta transformation.** See this entry.

start and stop bits the logic pulses of defined height and duration that indicate the onset and end of a stream of digital data.

starter 1. an additional electrode in a GAS-FILLED TUBE that starts IONIZATION. 2. the DEVICE that generates a VOLTAGE impulse to start the discharge in a fluorescent tube.

starting current the current flowing during starting conditions. The term usually denotes the heater current of a MAGNETRON which can be reduced when oscillation starts.

state the condition of output or signal level of a device, or level of any measurable quantity relating to the device. For many electronic devices, the state of the device will denote the voltage level at the output. Digital devices will allow only two states: 0 and 1.

static a type of radio or television INTERFERENCE. Static is produced by electrostatic discharges and is worst in the presence of thunder.

static characteristic a graph-plot of one quantity against another with all other variables held at constant values, such as a plot of transistor output current against input voltage, with collector voltage held constant. In a *dynamic characteristic* the effect of varying collector voltage would have to be considered, as this would then represent the changes in a working transistor.

static memory any form of memory that stores information without the need to rewrite the data at intervals (see REFRESH). Static memory may be VOLATILE like STATIC RAM or nonvolatile, like MAGNETIC CORE (2) storage. Neither type needs signals to maintain storage as would be the case with a DYNAMIC RAM memory.

static RAM a form of semiconductor MEMORY based on FLIP-FLOPS. Each flip-flop stores one bit of data, but only one output of each flip-flop is used. The same value of current will pass through each flip-flop whether a 1 or a 0 is being stored, so power consumption and the number of devices are greater for this type of memory than for a DYNAMIC RAM memory.

stationary wave see STANDING WAVE.

stator the stationary part, usually a magnet, of a motor or generator. Compare ROTOR.

steady state the state of any system after transient oscillations or other initial disturbances due to switching on have died away.

steerable antenna an antenna whose main LOBE can be rotated. This can be done either by moving the whole antenna structure, as is done with a radar dish, or by altering the phases of signals to separate components of the antenna array.

steering diode a DIODE used as part of a FLIP-FLOP circuit to permit TOGGLING. For a conventional flip-flop, two steering diodes are needed. The diodes are biased by the collector voltages of the transistor, so that a voltage input pulse of the correct amplitude can pass through only one diode. This will be the diode that is connected to the input of the shut-off transistor. As a result, the pulse switches the state of the flip-flop. After the changeover, the bias on the diodes will have changed also, so the state will reverse on the next pulse again. See Fig. 90.

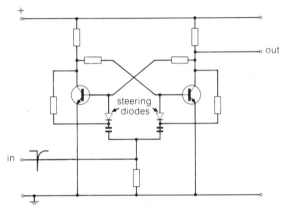

FIG. 90. **Steering diodes.** The steering diodes used in a bistable transistor circuit.

step-down transformer a TRANSFORMER connected so that its signal output is of lower voltage than its input. This implies that the SECONDARY WINDING has fewer turns than the PRIMARY WINDING. Compare STEP-UP TRANSFORMER.

step function a sharp change of voltage from one level to another, equivalent to half a SQUARE WAVE.

stepping relay a type of RELAY once used in telephone exchanges. The relay had a contact arm that could touch several other contacts in sequence as the arm rotated.

step-recovery diode see CHARGE-STORAGE DIODE.

step-up transformer a TRANSFORMER whose signal output is greater than its input. The SECONDARY WINDING contains more turns than the PRIMARY WINDING. Compare STEP-DOWN TRANS-FORMER.

stereo amplifier an audio AMPLIFIER having two channels. The channels must be separate, allowing virtually no CROSSTALK. They must, however, use ganged volume, treble, and bass control (see GANGED CIRCUITS).

stereophonic or **stereo** (of a sound-reproducing system) using more than one independent channel to produce the effect of sound that appears to come from an extended source, rather than from a single LOUDSPEAKER. Compare MONOPHONIC.

stereo receiver a radio receiver or TUNER that can make use of stereo transmissions. The FM signal is demodulated to recover the L + R signal, where L and R mean the left- and right-channel signals. At the same time, the sideband of the SUBCARRIER of the L − R signal is amplified, the carrier regenerated, and the L − R signal recovered by DEMODULATION. The L + R and L − R signals are then combined so as to separate the L and R signals.

stereoscopic (of an imaging system) producing an optical image that appears to have three dimensions, as in a LASER hologram.

stereo system a combined radio, cassette player, and phonograph, with amplifiers and loudspeakers.

stereo tape recorder a tape recorder that uses separate amplifying systems to record and replay stereo signals on two separate tracks of tape. A stereo record/replay head is needed, with two separate magnetic circuits, gaps, and coils.

stop band see REJECTION BAND.

stopper a RESISTOR used to suppress OSCILLATIONS. The stopper is connected close to a terminal of an active device. Its action is to add damping (see DAMPED) in an unintentionally oscillating circuit. See also BASE STOPPER.

stopping potential the reverse voltage needed to prevent an electron from landing on a surface. The stopping potential required on an ANODE measures the energy of the electrons liberated by light from a PHOTOCATHODE.

storage oscilloscope an oscilloscope that can retain a trace of a transient input, such as a short pulse. This was originally done using a direct-view STORAGE TUBE but is now performed by using digital storage.

storage time the time for which data can be stored before requiring to be refreshed. See REFRESH, DYNAMIC MEMORY.

storage tube or **charge-storage tube** a form of CATHODE-RAY TUBE that uses charge storage to retain its trace. Most types of TV CAMERA TUBES are storage tubes in this sense, but the term is usually reserved to denote direct-view display storage tubes. These tubes, developed in the 1950s, used metal mesh coated with an insulator and written with positive charges by an electron beam, using SECONDARY EMISSION to charge the insulator. The pattern of charge then affected the passage of an overall FLOOD GUN beam through the apertures in the mesh. This type of tube, the *direct-view storage tube (DVST)*, is still useful for some radar and oscilloscope applications, but has been superseded for most purposes by use of digital storage of waveforms.

strain gauge a method of measuring mechanical strain by electrical methods. The most useful methods are the RESISTANCE STRAIN GAUGE and the semiconductor strain gauge. See ELASTORESISTANCE.

strapping a technique for stabilizing MAGNETRON operating frequency. Strapping means the connecting of alternate cavities so that they share signals in the correct phase.

stray capacitance the CAPACITANCE that exists between parts of a circuit but is not intentionally designed into the circuit. This consists of the capacitance of connectors, wires, leads, and components themselves, both to ground and to other components. Stray capacitance is unwanted, but cannot be forgotten. It causes unwanted resonances, sometimes oscillations, and is responsible for the loss of gain at high frequencies in wideband amplifiers.

striking potential the POTENTIAL DIFFERENCE between two electrodes that is needed to start a gas discharge. See also STARTER (1).

strip core a magnetic core for an inductor that is made by winding a ribbon of SOFT MAGNETIC MATERIAL. Using a ribbon rather than a LAMINATED CORE makes it possible to use a material that is highly ANISOTROPIC magnetically.

strip line a form of TRANSMISSION LINE using a broad conductor, an insulating strip, and a narrow conductor.

strobe 1. a gating pulse for a signal. See STROBED DISPLAY. 2. See MOIRÉ PATTERN.

strobed display a display, usually of the SEVEN-SEGMENT type, that is displayed section by section. The display is operated by ENABLING (strobe) pulses that connect the anode voltage, thus illuminating each section of the display in turn. The rate of pulsing is such that all the units appear to be operating together.

Strobing enables all the connections to the segments to be taken to seven inputs irrespective of the number of digits in the display. When a strobe pulse is applied, the data for the correct unit are applied to the inputs. For example, when the tens unit is strobed, the data applied are for the tens figure. The data will also affect the bars of all the other units, but only the tens unit is visible because of the strobing.

strobe pulse a PULSE used to select one of a number of units.

stub a short length of TRANSMISSION LINE used for MATCHING sections. The stub may be either short-circuited or open-circuited at its far end. It is connected to the main line at a point where matching is needed, such as where the line joins to a device. The stub is then adjusted by cutting or by using a tuning capacitor connected to the end so as to achieve the minimum STANDING-WAVE RATIO.

subcarrier a second CARRIER (1) modulated onto a main carrier. See COLOR TELEVISION, PAL, STEREO RECEIVER.

subharmonic a frequency that is an integer fraction (½, ¼, etc.) of a FUNDAMENTAL FREQUENCY. Compare HARMONIC.

subsonic or **infrasonic** (of frequencies) being below the range of audible sound frequencies, that is, less than 20 Hz.

substandard an instrument that is a SECONDARY STANDARD.

substrate the material of a chip on which and of which an INTE-GRATED CIRCUIT is constructed, applied mainly to the main part of the semiconductor chip on which integrated circuit components and connections are made.

sunspot cycle the periodic changes of conditions in the outer layers of the sun. These cause considerable changes in the IONIZA-TION of layers in the IONOSPHERE thus affecting radio reception.

superconductivity a complete loss of electrical RESISTANCE. This occurs in many metals at temperatures close to absolute zero as well as in other substances at temperatures well above absolute zero. See also JOSEPHSON EFFECT.

superheterodyne receiver the most important type of RADIO RECEIVER circuit for virtually all usable frequencies. A superheterodyne receiver is based on the principle of changing the frequency of received signals to a lower frequency, the INTERMEDIATE FREQUENCY at which most of the amplification is carried out. This ensures that any unintended FEEDBACK such as is inevitable in the early tuned stages of a receiver, will not be at a frequency that causes oscillation. In addition, the main amplification can be carried out at a fixed frequency, affording greater gain and bandwidth than would be possible if a variable frequency had to be amplified.

The intermediate frequency is obtained by mixing the incoming frequency with the signal from a LOCAL OSCILLATOR. The oscillator-tuned circuit is adjustable, and the tuning of the oscillator is ganged (see GANGED CIRCUITS) to the tuning of the mixer input stage, and to any tuned premixer stage. In this way, the frequency of the oscillator is kept in step with that of the input signal so as to produce the correct intermediate frequency and do so consistently.

By convention, the oscillator frequency is higher than that of the incoming signal. See Fig. 91.

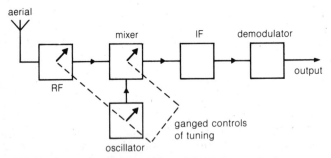

FIG. 91. **Superheterodyne receiver.** A block diagram for the superheterodyne receiver principle.

Supermalloy, *Trademark.* a magnetic nickel-iron alloy. Supermalloy contains 5% molybdenum and has very high permeability and low saturation field.

Supermendur *Trademark.* a magnetic cobalt-iron alloy. Supermendur contains vanadium and has very high saturation field strength.

superposition theorem or **principle of superposition** a useful electrical theorem for LINEAR NETWORKS of PASSIVE COMPONENTS. In a linear network containing several sources of voltage or current, the effect of each voltage or current at a selected point can be found separately, and these effects can be added to find the effect of all the voltages or currents at that point.

Suppose, for example, that a network contains two signal voltages. To find the voltage across one resistor in the network, imagine one signal source replaced by its internal resistance. Calculate the voltage across the resistor that is due to the remaining signal. Now reverse the situation, replacing the first signal but removing the second and substituting its internal resistance. Calculate the voltage across the resistor due to the second signal. Now add these voltages. The result is the voltage that would be caused by the two signal sources. The use of the superposition theorem is often simpler and quicker than methods based on KIRCHOFF'S LAWS.

superregeneration a method of operating a RECEIVER circuit for exceptionally high SENSITIVITY. The circuit is regenerative, meaning that positive feedback is used. The amount of positive feedback is so high that the circuit oscillates, but the circuit is arranged as a SQUEGGING oscillator, with the squegging permitted at an ultrasonic frequency and suppressed by the incoming signal. Superregenerative receivers can achieve astonishingly high gain, permitting use of two-transistor receivers, for example, but with poor selectivity. In addition, the radiation they set up by their oscillation interferes with other receivers around them, so superregeneration is used only in receivers for emergency use that must be of minimal size and power consumption. See REGENERATIVE RECEIVER.

suppressed-carrier transmission a system of TRANSMISSION and reception of AMPLITUDE MODULATION. In the suppressed-carrier system, the carrier frequency is suppressed or filtered to very low amplitude at the transmitter. Since half of the power of a conventional AM transmitter is used to transmit the carrier, and all the information lies in the SIDEBANDS suppression of the carrier greatly improves the efficiency of transmission. This can be increased still further if only one sideband is transmitted. The missing carrier must be regenerated at the receiver for purposes of DEMODULATION. This can be done either by CRYSTAL CONTROL of both transmitter and receiver, or by transmitting a small carrier signal that can be filtered, amplified, and used to synchronize a local oscillator in the receiver.

suppressed-zero indicator an indicator in which the scale does not start at zero. A simple example is the voltmeter in a car that has a range of only about 10 to 14 V.

suppressor a DEVICE that reduces the radiation of INTERFER-ENCE. The term usually denotes series resistors fitted in automotive ignition systems. These damp out the VHF oscillations in the system. Other suppressors employ C-R networks, diodes, inductors, and other components to slow down sudden voltage changes and damp out oscillations.

surface acoustic wave (SAW) an ULTRASONIC wave along the surface of a QUARTZ CRYSTAL or other material. Surface acoustic waves are generated by crystal TRANSDUCERS and can be of very high frequencies. The wave can be filtered precisely by mechanical treatment of the surface along which it travels. Surface wave filters (SAW filters) are therefore used extensively wherever a purely electronic filter cannot provide sufficiently sharp band edges and/or amount of attenuation.

surface-barrier device a transistor that uses SCHOTTKY DIODE junctions in place of the normal p-n type of junction.

surface leakage a leakage current along the surface of a material rather than through it.

surface passivation the treatment of a semiconductor or other surface so it cannot be affected by its environment.

surface resistivity the RESISTANCE per unit area of a surface. The resistance is measured between two opposite sides of a square.

surface wave a CARRIER (1) that follows the ground contours between transmitter and receiver.

surge a large and abrupt change of VOLTAGE. Surges are produced by lightning striking power cables, cable faults, reflections on a transmission line, or switching transients.

surge arrester a DEVICE that opens a CIRCUIT when a SURGE occurs.

surge diverter a CIRCUIT that limits the effect of a SURGE by limiting the surge voltage. A SPARK GAP is a simple form of surge diverter.

surge impedance the IMPEDANCE of a TRANSMISSION LINE to a sudden step of voltage. The value is determined by the values of inductance and capacitance per meter of line.

sweep a combination of a RAMP and a FLYBACK.

sweep voltage the linear rate of change of VOLTAGE from a TIME-BASE circuit that is used to deflect an ELECTRON BEAM. For magnetically deflected cathode-ray tubes, *sweep current* is a more useful quantity.

swing a range of values of VOLTAGE or CURRENT used by a DEVICE. The output swing of an amplifier stage, for example, can be between 1 and 9 V when a 10 V supply is used.

swinging choke an INDUCTOR whose inductance is deliberately made NONLINEAR. The inductance value of a swinging choke decreases as current increases, so the filtering effect of the choke is less effective for high currents.

switch a method of making, breaking, or changing electrical connections. Mechanical switches use moving contacts to make connections, but switching can also be carried out by semiconductor devices, for example, SILICON-CONTROLLED SWITCHES, THYRISTORS, TRANSISTORS, FIELD-EFFECT TRANSISTORS, DIODES.

SWR see STANDING WAVE RATIO.

SWR meter a form of radio-frequency measuring bridge that can measure the amplitude of forward and reverse waves on a line. By doing so, it indicates the value of voltage STANDING WAVE RATIO.

sync see SYNCHRONIZING PULSE.

synchronizing pulse a PULSE used to start an action so that the action will be in time with the pulse. Sometimes shortened to *sync*. The source of a sync pulse is called a *clock generator*. Synchronizing pulses are essential in MICROPROCESSOR equipment and are used extensively in TELEVISION circuitry. See also RASTER.

315

synchronous (of a SIGNAL) being forced into step with another signal. When different circuits are fed with the same sync pulse, they will be acting synchronously. The action of a television SYN-CHRONIZING PULSE is to make the TIMEBASE of a TV receiver operate synchronously with respect to the transmitted signals. Synchronous action is also important in a computer system, in which virtually every waveform is synchronized with a main CLOCK PULSE. Compare ASYNCHRONOUS; see also SYNCHRONOUS COUNTER.

synchronous counter a set of FLIP-FLOPS arranged so as to act as a counter of PULSES that are fed to each flip-flop in the counter. Unlike the ASYNCHRONOUS COUNTER, in which each flip-flop is toggled (see TOGGLING) by the output of the previous flip-flop, the synchronous counter applies the input pulses to the clock termi-nals of each flip-flop, so that clocking occurs at the same time for each unit. The counter uses J-K FLIP-FLOPS and the counting logic is determined by the way in which the J and K inputs are con-nected by means of GATES.

synchronous demodulation see SYNCHRONOUS DETECTION.

synchronous detection or **synchronous demodulation** or (formerly) **homodyne detection** a form of DEMODULATION in which a modulated signal is mixed with a sine wave that is at carrier frequency. The modern synchronous detector is phase sen-sitive (see PHASE DISCRIMINATOR), so its output is affected by any difference between the phase of the incoming signal and the phase of the locally generated carrier.

synchronous gating a method of opening and closing signal paths in response to a SYNCHRONIZING PULSE.

synchronous logic the use in a LOGIC CIRCUIT of FLIP-FLOPS that are clocked (see CLOCK). The clock signal is independent of the inputs to the flip-flops, and the advantage of the system is that it avoids RACE HAZARDS by ensuring that all logic stages operate at precisely controlled times.

synchronous motor a type of AC motor in which the rotating speed is determined entirely by the power supply frequency. In general, the synchronous motor will either run at its synchronous speed or not at all. Some types, such as those used in AC-powered electric clocks, are not self-starting.

sync separator a circuit in a TELEVISION RECEIVER used to separate the two SYNCHRONIZING PULSES from the VIDEO SIGNAL and from each other. The sync separator first separates the synchronizing signals from the video waveform, using a biased amplifier stage. The two sync signals then are separated from each other. A differentiator circuit is used to give a sharp spike for each sync pulse, thus providing a horizontal (line) synchronizing pulse (see RASTER).

The differentiator will also give an output for each of the field sync pulses, but this is unimportant. An integrating circuit provides a negligible output for each line sync pulse, but the more closely spaced field sync pulses cause a rise in voltage, which can be used to synchronize the vertical (field) timebase generator. See Fig. 92. See also RASTER.

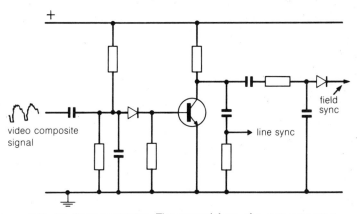

FIG. 92. **Sync separator.** The general form of a sync-separator STAGE before integrated circuits took over the functions.

synthesis the construction of a WAVEFORM or a desired response from standardized components (see STANDARDIZATION). An example is *filter synthesis* the technique of designing filters of desired characteristics from standard values of inductors, capacitors, and resistors.

synthesizer a DEVICE that uses electronic CIRCUITS for generating

317

SOUNDS. Synthesizers are based on the generation of SQUARE WAVES in the audio range. These waves are rich in HARMONICS. FILTERS are used to modify the shape of the waves, and modulators, limiters, clippers, and other devices are used to shape the sound wave and form an envelope of waves. This makes it possible for a synthesizer to imitate the sounds of instruments and, more important, to create new types of notes and effects.

Many synthesizers can be connected to computers using the MIDI (musical-instrument digital interface) system to permit a degree of control difficult to achieve manually.

system a set of interacting parts that make up a working entity. The term can denote a circuit, device, a construction such as a radio receiver or a radar station, or a large assembly such as a steel mill. Anything can be treated as a system if it has definable inputs, outputs, and control actions.

T

---

tachogenerator a miniature electrical GENERATOR for measuring rotational speed. The tachogenerator is designed to give a linear voltage output at all speeds of shaft rotation. This permits rotating components to be included in a FEEDBACK loop, with the tachogenerator providing the feedback signal proportional to speed.

tachometer any form of angular speed sensor. Electronic tachometers can use revolving magnets with static pickup coils, light beams that are interrupted by a revolving object such as a slotted disk or a TACHOGENERATOR.

tap changer or **ratio adjuster** a device for changing a TRANSFORMER ratio. The term usually denotes a switching type of changer that can be used while the transformer is energized. An alternative to tap changing is to use a fixed ratio transformer in conjunction with a variable transformer such as a VARIAC.

tape see MAGNETIC TAPE, PAPER TAPE.

tape hiss a NOISE heard when a magnetic tape is replayed. The noise is concentrated in the high audio frequencies, so it sounds like a hiss rather than a rumble or boom. Hiss can be minimized by using a wide section of tape for recording, and moving the tape rapidly past the head. Where these solutions are not available, noise-reduction systems that rely on companding (see COMPANDER) must be used. See also DOLBY SYSTEM, DBX, RECORDING OF SOUND.

tape recording see RECORDING OF SOUND.

tapped-coil oscillator see HARTLEY OSCILLATOR.

tapping a connection to a component that is not to either end terminal. RESISTORS and INDUCTORS are the most common tapped components.

target an ELECTRODE at which an ELECTRON BEAM is aimed. The

319

electrode is used in 'TV CAMERA TUBES and other charge STORAGE
TUBES. The target of an X-RAY TUBE is also called an *anticathode*.

target voltage 1. the voltage between the CATHODE of an electron
gun and the TARGET of a TV camera tube. This voltage is normally
low, in the order of 10 to 30 V. **2.** the voltage between the filament
and the target of an x-ray tube. In this case, the voltage will be very
high, approximately 100 kV.

Tchebycheff filter a filter that allows limited passband ripple to
occur but gives a steeper transfer function in the transition region
between PASSBAND and STOP BAND. The amplitude response of
this type of filter is given by

$$\frac{V_{\text{out}}}{V_{\text{in}}} = \frac{1}{[1 + \epsilon^2 C_n{}^2 (f/f_c)]^{\frac{1}{2}}}$$

where C is a polynomial of degree n, ϵ is a constant, f_c is the
corner frequency, and f the frequency.

tearing the breakup of a TV image because of faulty synchronization of timebase waveform (see RASTER). The most common form
is line tearing, in which the picture seems to be torn horizontally.
In color receivers, faulty color synchronization can cause color
tearing (see HANOVER BARS).

Teflon see PTFE.

telecommunications the transmission and reception of signals in
electrical form by wire or by electromagnetic wave.

telegraphy communication by signal codes along wires or by
radio. For message telegraphy, MORSE CODE is still in use, but the
increasing use of computers has led to acceptance of BINARY CODE
systems.

telemetry the remote measurement of physical quantities by
TRANSDUCERS and some form of transmitter, for example, radio.
See also RADIOSONDE.

telephony any method of transmitting audio-frequency signals,
either by wire or by radio.

teleprinter an electrically operated typewriter that is controlled
by digital signals. The term is now used only to denote an obsolete

type of electromechanical device that has mainly been superseded by DOT-MATRIX and daisy wheel printers.

teletext a television information service providing data consisting of digits, letters, and some standardized graphics symbols. These are coded, using the digital ASCII code system, and transmitted along with the TV signal. The transmission of teletext takes place during the field-flyback period, and the signals are decoded, stored, assembled, and displayed by a suitable receiver or adaptor. The signals can also be received by an adaptor connected to a suitable computer.

television (TV) a method of transmitting sound and animated vision signals. The TELEVISION CAMERA uses light-sensitive CAMERA TUBES on the face of which the image is projected by lenses. The tubes are scanned, so that data on brightness (and color) are extracted serially to form the VIDEO SIGNAL. To this are added BLANKING and SYNCHRONIZING PULSES and the modulated color SUBCARRIER signals, to form a COMPOSITE video signal.

Sound signals from microphones are amplified separately. The two sets of signals are modulated onto separate UHF carriers, with the SOUND CARRIER above the vision carrier frequency, and frequency modulated. See also TELEVISION RECEIVER, COLOR TELEVISION.

television camera a camera whose input is an optical image and whose output is a VIDEO waveform. The camera consists of an optical system, CAMERA TUBES and PREAMPLIFIER stages. For a color camera, at least three pickup tubes are used, and the optical system is very elaborate. It must consist of a main lens, often of the zoom variety, in which the image path is split into as many sections as there are camera tubes, so that each tube has an image focused on its face. By using dichroic prisms or mirrors, these images will be in the primary colors of blue, red, and green—some designs use a black/white image also.

The pickup tubes are almost always VIDICONS and their output is video signals. The electron beams of the vidicons are deflected by TIMEBASE signals, all of which are synchronized to a master generator in a camera control unit, often separate from the camera. BLANKING pulses are used to suppress the output from the vidicons during the FLYBACK time of the timebases. The separate video signals are amplified in a preamplifier, and any frequency corrections and corrections for other distortions such as aperture distortion are made at this point.

The convention is that the video waveforms are amplified to a peak-to-peak amplitude of 1 volt. The video signals are then transmitted from a low output impedance current amplifier through a cable to the camera control unit. In this unit, further correction is made, sync signals are added, and the color signals are processed. This consists of forming a LUMINANCE signal (black/white) from the color signals if no separate luminance signal has been generated. The color signals, in pairs, are then mixed to form signals that are modulated onto SUBCARRIERS using in-phase and quadrature phase modulators. The COLOR BURST of subcarrier frequency is added to the BACK PORCH of the line sync pulse, and the modulated subcarrier signals are added to the video signal, with the carrier itself suppressed.

television receiver a device containing circuitry for receiving TV signals and producing an image. The receiver follows the pattern of a SUPERHETERODYNE RECEIVER system, and a broadband IF signal is obtained, containing both video and sound signals. The vision demodulator produces the main LUMINANCE signal, and enables the color signals, which are modulated onto a SUBCAR-RIER, to be separated off along with the INTERCARRIER SOUND signals. The sound signal is separately amplified, demodulated, and passed to the audio stages. The COLOR BURST is used to synchronize a local color oscillator, and the output of this is used in two synchronous detectors (see SYNCHRONOUS DETECTION) to obtain the two color difference signals. These signals are then mixed with the luminance signal in the correct proportions to recover the separate red, green, and blue signals for the guns of the color cathode-ray tube.

Picture synchronization is achieved by separating off the mixed sync pulses from the luminance signal in a sync separator that also separates the sync pulses from each other. These pulses are used to synchronize the TIMEBASE circuits that generate the linear sweep (ramp) currents for the deflection coils. The line-sweep current must be switched off rapidly to achieve a fast flyback, and the voltage produced by this rapid change of current in the line output transformer is stepped up in a WINDING rectified by a VOLTAGE MULTIPLIER and used to provide the accelerating voltage (extra-high voltage) for the picture tube.

telex a system in TELEGRAPHY for digitally coded signals. Telex users can type or receive messages on a TELEPRINTER or other terminals, and adaptors can be used to enable computers to be linked to the network.

temperature coefficient a measure of the effect of changing temperature on a physical quantity. The coefficient is defined as the fractional change of quantity per unit temperature change. The temperature coefficient may be positive (increasing quantity for increasing temperature) or negative (decreasing quantity for increasing temperature). The most important temperature coefficients in electronics are of resistivity (or resistance) and of capacitance. Most ohmic conductors have a positive temperature coefficient of resistivity or resistance. Semiconductors generally have a negative coefficient. Both positive and negative temperature coefficient capacitors can be obtained, and oscillator circuits often use a mixture of capacitor types in order to balance out the effect of temperature changes on the tuned circuit.

TEM wave a transverse electromagnetic wave, that is, a wave in which the electric and the magnetic fields are perpendicular to each other and also to the direction of PROPAGATION. This is the type of wave that is propagated through space and along TRANS-MISSION LINES. It cannot be propagated along a WAVEGUIDE.

terminal 1. a place to which leads can be attached into or out of a CIRCUIT. 2. the remote keyboard and display unit of a computer system.

terminal impedance the IMPEDANCE at the end of a TRANSMIS-SION LINE.

termination a MATCHING load for a TRANSMISSION LINE that prevents reflections.

tertiary winding an additional WINDING on a TRANSFORMER. The tertiary winding can be used to supply a FEEDBACK signal, provide a voltage level that differs from the level on the SECONDARY WINDING or ensure insulation. For example, a well-insulated tertiary winding is used on a transformer for an oscilloscope to supply the heater voltage for the CATHODE-RAY TUBE because the cathode may be operating at a large negative voltage. A tertiary winding that is electrically connected to another winding is often called an *overwind*. See also PRIMARY WINDING.

test pattern a definition and shading pattern used to check a television system. The chart contains fine, closely spaced lines and shaded blocks that should be distinguishable from each other, even by home TELEVISION RECEIVERS.

tetrode see VACUUM TUBE.

TE wave a transverse electric wave in a WAVEGUIDE in which the electric field is POLARIZED (1) at right angles to the direction of propagation, but with a magnetic field component whose polarization varies with distance traveled along the waveguide.

THD *abbreviation for* total harmonic distortion, the sum of the ROOT-MEAN-SQUARE amplitudes of all HARMONICS that are generated by an amplifier whose input is a pure sine wave. The THD is usually expressed as a percentage at full rated power, or shown as a graph plotted against power. For ordinary purposes, a THD of up to 10% is tolerated. Hi-fi sound systems deal with THD amounts in the region of 0.1% or less.

thermal breakdown the complete loss of correct action of a component due to overheating. The overheating can in turn be caused by RUNAWAY.

thermal conductivity a quantity that measures the ability of a material to conduct heat through its bulk. The thermal conductivity coefficient for a material is defined as the quantity of heat transmitted due to a unit temperature gradient, in unit time under steady conditions and in a direction normal to a surface of unit area. The coefficient is independent of sample size and shape. See also THERMAL RESISTANCE.

thermal imaging the process of producing an image of an object from its heat radiation. Thermal-imaging cameras use TV techniques with infrared-sensitive CAMERA TUBES. Thermal imaging is used for military purposes and in crime prevention and detection. It is also used in medicine, because cancerous cells are often at a slightly higher temperature than healthy tissue and show up differently on the thermal display.

thermal instrument any type of measuring instrument that makes use of heating. See HOT-WIRE METER.

thermal noise or **Johnson noise** the electrical NOISE signal caused by the thermal agitation of current CARRIERS (1). In any conductor, the current carriers vibrate as well as move along a current path. The vibration constitutes a noise signal, and its amplitude is proportional to the square root of (absolute) temperature. If very low noise is required, an amplifier may have to be operated at very low temperature.

thermal resistance a measure of the RESISTANCE of a material to heat flow. The practical units used in electronics are degrees Celsius per watt, and the measurement expresses the change of temperature caused by the dissipation of each watt of power. The thermal resistance of a collector junction to the case of a transistor is an important PARAMETER for a power transistor, because it determines the maximum power level at which the transistor can be used safely.

thermal runaway the loss of control over a device at high temperature. See RUNAWAY.

thermionic of or relating to any DEVICE that uses ELECTRONS emitted from a hot surface.

thermionic cathode a hot electron-emitting surface. This may be a tungsten or tungsten/thorium wire that is heated to temperatures of about 2,500 K by passing direct current through it. A cathode of this type is called directly-heated, that is, a FILAMENT.

Indirectly heated cathodes are used in CATHODE-RAY TUBES, CAMERA TUBES and some small VACUUM TUBES. These use cathodes that are hollow nickel tubes coated with a mixture of oxides of barium, strontium, and calcium. This material emits electrons copiously at comparatively low temperatures, about 1,000 K. The cathode is heated by an insulated molybdenum spiral filament inside the nickel tube.

The disadvantage of using such materials is that the cathode material is easily damaged, particularly by sparking or by the electrostatic removal of the coating in large fields, and it tends to liberate gas when used for high currents. For these reasons, transmitting tubes use tungsten or thoriated tungsten filaments.

thermistor a nonohmic resistor (see OHMIC.). Thermistors are made of semiconducting materials, and most types have large NEGATIVE TEMPERATURE COEFFICIENTS. They are used for temperature measurement, temperature warning, and temperature compensation. Thermistors are also used in the FEEDBACK circuits of R-C oscillators to stabilize the amplitude of oscillation by increasing the amount of NEGATIVE FEEDBACK when the amplitude of oscillation becomes excessive.

thermoammeter a current METER that uses the heating effect of CURRENT. The thermoammeter uses either the HOT-WIRE METER principle, or a THERMOCOUPLE heated by current through a strip of resistive material. Thermoammeters are imprecise, but they often are the only way of measuring a radio-frequency current. They can also be calibrated in terms of true RMS values.

thermocouple a JUNCTION (1) between two different metals. Two thermocouples connected in series will generate a voltage if the junctions are at different temperatures. The voltages are of millivolt level, but comparatively large currents can be passed. Thermocouples are used extensively for measuring temperature because the sensor (junction) can be remote from the meter at which the reading is taken. See also THERMOELECTRIC EFFECT.

thermoelectric effect a term applied to a set of discoveries by Thomas Seebeck (1770—1831), Jean Peltier (1785—1845), and Lord Kelvin (1824—1907). The *Seebeck effect* is the action of the THERMOCOUPLE. The *Peltier effect* is the opposite, the heating of one junction and the cooling of the other when current is forced to flow through two junctions. By using metal/semiconductor junctions, large temperature changes can be obtained, so refrigeration by Peltier junction is possible. The *Kelvin effect* is a voltage in a single conductor caused by temperature differences that affect the electron ENERGY BANDS.

thermography a form of THERMAL IMAGING that produces a permanent image on paper or on film.

thermojunction a single junction of a THERMOCOUPLE. The favorite pairs of materials for junctions include copper and constantan, iron and constantan, and platinum and rhodium.

thermopile a set of THERMOCOUPLES wired in series or in series-parallel to form a source of ELECTROMOTIVE FORCE comparable to that of a chemical battery.

thermostat a temperature-operated SWITCH. The usual thermostat principle uses a BIMETALLIC STRIP carrying a switch contact.

The device suffers from considerable HYSTERESIS, which limits its usefulness as a method of maintaining a constant temperature. Thermostats now are being superseded by THERMISTORS often as part of a MICROPROCESSOR system.

Thevenin's theorem a method of analyzing linear resistance networks. The theorem states that any two-terminal LINEAR NETWORK that contains sources of voltage and resistances can be represented by an equivalent single voltage source and series RESISTANCE. With Thevenin's theorem, an EQUIVALENT CIRCUIT can be drawn up for any linear network. See also NORTON'S THEOREM.

thick-film circuit a type of miniature PRINTED CIRCUIT BOARD but with PASSIVE COMPONENTS formed in the connections. The circuit is formed by photolithographic techniques, using metal films for connections and for resistors, with capacitors formed from metal and oxide layers. INTEGRATED CIRCUITS can be bonded to a thick-film circuit by compression welding to bonding pads. Compare THIN-FILM CIRCUIT.

thin-film circuit a form of circuit that uses sputtered (see SPUTTERING) or vacuum-coated films rather than thick films. Compare THICK-FILM CIRCUIT.

thoriated tungsten a thorium-tungsten alloy material, used for directly heated CATHODES in transmitting tubes. The presence of thorium enables electron EMISSION at considerably lower temperatures than tungsten alone. When a thoriated tungsten cathode begins losing emission, it often can be regenerated by running the FILAMENT for a short time at a higher temperature than normal, thus allowing more thorium to diffuse to the surface.

three-phase supply a supply of AC along three separate lines, with 120 degrees between the voltage phases of any two lines.

three-state logic a LOGIC CIRCUIT arrangement for INTEGRATED CIRCUITS. In a three-state logic terminal, the voltage can be at logic 0, logic 1, or FLOATING. In the floating state, the voltage can be changed by external voltages without damaging the internal circuits of the chip. See also TRI-STATE DEVICE.

threshold a point at which a cause begins to produce an effect. The threshold frequency for the PHOTOELECTRIC EFFECT is the light frequency for which EMISSION becomes detectable. The threshold voltage for a conducting device is the voltage at which the device begins to pass current. The threshold signal for a receiver with SQUELCH is the signal that overcomes the squelch, thus enabling the audio section of the receiver.

thyratron a GAS-FILLED TUBE that acts as a RELAY. Thyratrons have largely been replaced by their semiconductor equivalent, the THYRISTOR.

thyristor or (formerly) **silicon-controlled rectifier (SCR)** a SEMICONDUCTOR pnpn device. The thyristor has three electrodes: ANODE, CATHODE, and GATE (1). When a voltage of either polarity is applied between the anode and the cathode, current will flow between the anode and the cathode only when the anode is positive with respect to the cathode, and when the gate has been triggered. The TRIGGER PULSE is applied to the gate at the time when the anode is positive. The effect of the trigger pulse is to make the device conduct heavily. Conduction continues until the anode voltage drops to almost the cathode voltage level, or until the current falls to a very low value. Thyristors have almost completely replaced THYRATRONS for most purposes.

tight coupling a type of tuned COUPLING in which the COUPLING COEFFICIENT is large, so that a large fraction of the input signal is transferred to the output. Tight coupling is efficient in terms of signal transfer and can provide optimum signal transfer conditions if the coupling is critical (see CRITICAL COUPLING). If the coupling passes the critical point, then OVERCOUPLING will result, causing the bandwidth to increase considerably. It will also cause the shape of the transfer characteristic (output plotted against input) to develop a trough between two peaks. Compare LOOSE COUPLING.

timebase a CIRCUIT that produces a WAVEFORM for ELECTRON BEAM deflection. The most common type is the linear RAMP (or SWEEP) circuit, which produces a SAWTOOTH WAVEFORM. Timebase generators use a combination of a SQUARE-WAVE generator and an INTEGRATOR to produce the linear ramp (or sweep), and some form of switching circuit to produce a rapid FLYBACK. By separating the square-wave generator portion from the integrator, it is possible to control the frequency and slope of the timebase sweep separately. The square-wave generator may be a MONOSTA-

BLE in which case the timebase is a triggered one, producing a sweep voltage only when a TRIGGER PULSE arrives.

time constant a measure of the time required to change the VOLTAGE across a CAPACITOR or INDUCTOR. For a capacitor-resistor circuit, the time constant is given (in seconds) by $C \times R$ where C is in farads and R in ohms. For an inductor-resistor circuit, the time constant is L/R with L in henries and R in ohms. The time obtained from these calculations is the time needed for the voltage across the capacitor or the current through an inductor to decrease to $1/e$ or 36.8% of the starting value. Time constant values are particularly useful in estimating the effect of a CR or LR network on a square wave.

time delay or **time lag** the time between a signal being applied to the INPUT (1) and appearing at the OUTPUT of a signal.

time-delay switch a SWITCH that operates after a specified time delay. In equipment that makes use of indirectly heated CATH-ODES a time-delay switch is often used to ensure that high voltage is not applied until the cathode has warmed up.

time discriminator a demodulator for pulse-time modulated signals. The time discriminator generates a signal whose amplitude is proportional to the time between pulses. A simple form of time discriminator is a TIMEBASE that is triggered by one pulse and terminated by the next.

time-division multiplexer a SYSTEM by which a cable or radio channel can be shared by several signals. Each signal is sampled, and the sample amplitudes are transmitted in sequence. At the receiver, the same rate of sampling has to be used to reconstitute the signals correctly. SYNCHRONIZING PULSES can be transmitted at intervals to keep the transmitter and the receiver in step.

time lag see TIME DELAY.

time-sharing the sharing of the use of a computer by several users. By interleaving the signals to and from the main computer, each user can be served, generally without unnoticeable time delays. The system becomes unworkable if all users simultaneously need the full speed of the computer.

time switch a form of CLOCK either electromechanical or digital quartz, that will switch circuits on or off.

time to trip the circuit-breaking time for a relay. This is the time between activating the coil or deactivating it and breaking the circuit that is operated by the relay.

T-junction see HYBRID JUNCTION.

TM wave a transverse magnetic wave in a WAVEGUIDE in which the magnetic field is POLARIZED (1) at right angles to the direction of propagation, but with an electric field component whose polarization varies with distance traveled along the waveguide.

toggling the switching of STATE as of a FLIP-FLOP in which each PULSE at the INPUT (1) causes the OUTPUT to change state. A toggling circuit is a divide-by-two COUNTER since two complete pulses at the input produce one complete pulse at the output. A series of connected toggling circuits can be used to provide divide-by 4, 8, 16, and so on counters.

tolerance the amount of variation about a nominal value. Most passive components in electronic circuits are specified with quite wide tolerances, up to 20%. The tolerance of value for electrolytic capacitors can be as high as 50% of nominal value, and semiconductors also have large tolerance values.

The problem of wide tolerances is dealt with in analog circuits by using NEGATIVE FEEDBACK circuits in which only a few components in the feedback loop need be of close tolerance. In digital circuits, tolerances have little effect on the action of correctly designed circuits.

tone arm the arm that holds and guides the PICKUP of a phonograph. The design of a tone arm is complicated by conflicting requirements. One is the need to hold the pickup head rigid, while allowing it free movement to follow the record track without any vibrational resonance that would cause peaks in the response.

tone control a control over the FREQUENCY response of an audio AMPLIFIER. Simple radio receivers sometimes incorporate a treble cut control, but good quality sound reproduction demands a more comprehensive system of tone controls. These should include the actions of boosting or cutting BASS response, with a choice of CROSSOVER FREQUENCY and a similar variable boost or cut for TREBLE with a choice of crossover.

toroidal inductor an INDUCTOR wound around a ring-shaped core. The toroidal shape of a MAGNETIC CORE (1) is particularly efficient because it is easy to ensure that all its flux passes through the coils wound around it. Now that machines for winding toroidal coils are widely available, toroidal transformers are displacing the older LAMINATED CORE type.

total emission the total electron current from a CATHODE to all forms of GRID or ANODE.

totem-pole circuit a SINGLE-ENDED push-pull circuit. This is a popular type of transistor output stage for digital devices and audio amplifiers. In its simplest form, it consists of two COMPLEMENTARY TRANSISTORS connected in series. The same signal is taken to both bases, with a steady bias voltage superimposed. Each transistor then deals with half of the output signal. Since each transistor is being used in CLASS B the stage is efficient and, with NEGATIVE FEEDBACK and care about biasing, can produce quality output. See Fig. 93.

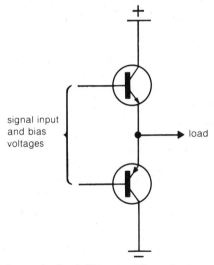

FIG. 93. **Totem-pole circuit.** The general form of a totem-pole circuit using complementary transistors. The output can sometimes be direct-coupled if balanced positive and negative supplies are used.

touch control a type of SWITCH operated by touching. The switch contains contact areas, and touching with the finger makes a high-resistance connection, which triggers a transistor circuit and operates a THYRISTOR. The touch portion of the circuit must be completely isolated from the thyristor circuit. Some touch switches provide safety by making use of the capacitance between an insulated contact and the body of the user.

trace the shape drawn by the SPOT (1) of light on the face of a cathode-ray tube.

track conducting line on a PRINTED CIRCUIT BOARD or a line of stored information on a TAPE or a DISK.

tracking 1. the maintenance of a fixed relationship between quantities. **2.** in a SUPERHETERODYNE RECEIVER the maintenance of correct tuning in the RF and oscillator circuits, so that the difference between their frequencies is constant (see also GANGED CIRCUITS). **3.** the locking of a RADAR beam onto a target in order to have the beam continue to follow the target to the extremes of the radar's range until it is reset. **4.** in a PHONOGRAPH, the arrangement of the TONE ARM so that the stylus is always correctly located in the groove of the disk. **5.** the formation of conducting tracks that will cause sparking on an insulator.

trailing edge the falling-voltage edge of a PULSE.

transconductance the ratio of change of output current to change of input voltage for an ACTIVE COMPONENT. Symbol: *Gm;* unit: usually milliamperes per volt. Also called *mutual conductance*. See also CONDUCTANCE.

transducer a CONVERTER of energy from one form to another. For electronics purposes, the transducers of interest are those that have either an electrical input (actuators) or output (sensors).

transductor see SATURABLE REACTOR.

transfer characteristic a graph plot of output against input for a device or circuit. The CHARACTERISTIC may be of current (output current/input current), voltage, or a hybrid such as output voltage/input current.

transfer current the current to the STARTER (1) of a gas-filled device.

transfer parameter a PARAMETER calculated from the slope of a TRANSFER CHARACTERISTIC.

transformer a general term for a pair of mutual inductors (see INDUCTANCE). For high-frequency signals, a transformer can take the form of two coils placed close to each other. For lower frequencies, MAGNETIC CORES (1) must be used to concentrate the FLUX and ensure that all the flux from one winding cuts the turns of the other winding. The winding to which a signal is applied is the PRIMARY WINDING. The winding from which a signal is taken is the SECONDARY WINDING. For an IDEAL TRANSFORMER the ratio of secondary voltage to primary voltage is the same as the ratio of secondary turns to primary turns.

transient a brief pulse of voltage or current.

transistor a three-electrode SEMICONDUCTOR device. The main types of transistors are the BIPOLAR TRANSISTOR and the FIELD-EFFECT TRANSISTOR (FET). In either type, a main current between two of the electrodes is controlled either by a much smaller current at the third electrode (in the bipolar transistor) or by a small voltage at the third electrode (FET). Both types make use of p-type and n-type semiconductor material, but they differ in the way in which the main current is conducted and controlled.

In the bipolar transistor, the main current between emitter and collector terminals flows through two JUNCTIONS (1) and is controlled by passing current into or out from the region between the junctions. In the field-effect transistor, the main current flow between source and drain terminals is along a path of one material type, which can be p-type or n-type.

The modern methods of fabrication of transistors are EPITAXY and ION IMPLANTATION and although DISCRETE devices are still manufactured in great numbers, most transistors are now formed as parts of INTEGRATED CIRCUITS.

transistor parameters the measurements that allow the action of a transistor to be represented in an EQUIVALENT CIRCUIT. Many sets of PARAMETERS were devised during the development of transistor technology, and many of these are no longer used. The almost universal adoption of silicon as a transistor material has meant that many of the old transistor parameters have values that can be ignored.

A useful equivalent circuit of a silicon bipolar transistor is one in which the transistor is considered as a current generator producing 40 mA per input volt for each milliamp of bias current. This leads to a figure of voltage gain of 40 × (with bias current in mA) for transistors operated with small signals in the audio frequency region. More detailed analysis is needed if the transistors are to be used in radio frequency or wideband circuits, and for switching circuit design, the large-signal equivalent circuit must be used.

transistor-transistor logic (TTL) a system of constructing gate and other logic circuits in INTEGRATED CIRCUIT form with BIPOLAR TRANSISTOR techniques. The gates use integrated transistors whose bases are connected through a resistor to the supply positive voltage terminal.

The input logic signals are to the EMITTERS (1) of these transistors. A logic 1 input, therefore, requires no current to or from the source. A logic 0, however, requires the source voltage to drop to zero, and to accept current from the integrated circuit input. (The driving circuit must be able to SINK(1) current.)

TTL circuits are fairly fast-acting, but require a considerable amount of supply current. The later Schottky devices make use of SCHOTTKY DIODES in circuits that permit input to the base of a transistor. These devices require much lower driving currents and are very fast in operation. See Fig. 94.

transition a sudden change from one STATE to another, usually of electron energy level inside an atom.

transit time the time an ELECTRON or HOLE takes to travel from one ELECTRODE to another in an ACTIVE COMPONENT. Transit time for a device may limit its ability to operate with high-frequency signals.

transmission the sending of SIGNALS from one place to another. Compare RECEPTION.

transmission line or **line** a CABLE or arrangement of cables for transmitting SIGNALS. It is important that a transmission line not radiate or reflect signals. If the signal at the input of the line is to reach the end of the line substantially unaffected, then the line must be matched (see MATCHING) to the impedances of the load and the source so that the STANDING-WAVE RATIO is as low as possible. Use of a balanced or a shielded line reduces radiation, and correct matching reduces reflection to a minimum. Loss of power

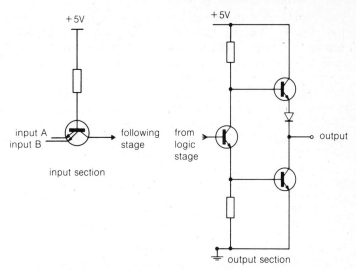

FIG. 94. **Transistor-transistor logic.** The basis of TTL circuits. The input is to a transistor with multiple emitters, and the output is from a type of totem-pole circuit.

on the line can be minimized by good construction and suitable materials.

transmission loss the ratio of power out to power in for a TRANSMISSION LINE or other system. The loss figure is often converted into decibels.

transmission primary a primary color for COLOR TELEVISION use. The transmission primaries are red, blue, and green. These are not the primary colors of paint. The primary light signals add to white, and the paint primaries add to black.

transmit/receive (TR) switch a gas-discharge device used in a RADAR system. The gas ionizes (see IONIZATION) during the transmission period, placing a short circuit across the receiver waveguide. This prevents the receiver from being paralyzed by the transmitted signal. Following transmission, the gas deionizes, and the receiver waveguide opens again. Compare ANTI-TRANSMIT/ RECEIVE SWITCH.

335

transmitted-carrier transmission a conventional amplitude-modulation system with full carrier amplitude present. See SUPPRESSED-CARRIER TRANSMISSION, VESTIGIAL SIDEBAND.

transmitter a device that sends out a carrier signal at radio frequency modulated by audio, video, or other waveforms. Compare RECEIVER.

transponder a DEVICE that receives a coded signal, the interrogating signal, from a distant TRANSMITTER and responds with another coded signal for recognition purposes. The transponder is the basis of the IFF system, a type of SECONDARY RADAR.

transposed line a LINE (1) or cable in which the positions of the conductors are interchanged at intervals to reduce any possibility of losses caused by asymmetry of capacitance or inductance.

transputer a complete computer, including the MICROPROCESSOR and its MEMORY fabricated on a single CHIP.

transverse wave any wave in which OSCILLATION occurs at right angles to the direction of PROPAGATION. All electromagnetic waves are transverse, but acoustic waves are often longitudinal, oscillating along the direction of propagation.

trap a type of FILTER that reduces the amplitude of a particular frequency, usually by means of a resonant tuned circuit. A trap is used in a color TV receiver, for example, to eliminate the color subcarrier signal from the luminance signal before the luminance signal is applied to the cathode-ray tube.

trapezium distortion a form of distortion of a TV image in which the normal rectangular shape is distorted into the shape of a trapezium.

traveling wave an electromagnetic wave (see ELECTROMAGNETIC RADIATION) traveling in space or along a TRANSMISSION LINE with no reflections. At any point along the direction of travel, the wave amplitude will vary with time at the wave frequency. Compare STANDING WAVE.

traveling-wave tube (TWT) a type of AMPLIFIER for MICROWAVES. The waves are forced to travel in a spiral path along a line inside a vacuum tube. A beam of electrons is made to travel down the axis of the tube at the same speed, with the result that the ELECTRON BEAM is modulated by the waves. The electron beam, which is now bunched (see BUNCHER) by the interaction, passes its power back to the wave at a later part of the spiral, so that the output power is much greater than the input power.

In a variation of the principle, the direction of the electron beam is opposite to the direction of the wave, constituting a BACK-WARD-WAVE TUBE. The traveling-wave tube is an important means of low-noise amplification at microwave frequencies, but it has been eclipsed to some extent in this role by the MASER.

treble the AUDIO FREQUENCIES in the range 2 kHz upward.

treble response a feature of an AMPLIFIER or SYSTEM that is measured by the amplitude of treble frequencies relative to lower frequencies. The standard frequency for response is often that of a 1 kHz wave, and the relative values of amplitude are usually expressed in decibels.

triac a type of two-way THYRISTOR. The triac uses an electrode (the gate) to trigger conduction between two other electrodes, usually labeled M1 and M2, since the terms anode and cathode are not appropriate. The conduction ceases when the current through the device falls to below the HOLDING CURRENT value or when the voltage between M1 and M2 drops below the holding value.

trickle charge a small amount of CURRENT supplied to a SECOND-ARY CELL in order to charge the cell slowly or maintain it in a fully charged state.

trigger level a critical level of VOLTAGE that must be exceeded by a TRIGGER PULSE in order to start an action.

trigger pulse a brief PULSE that will start an action. A trigger pulse will normally have a sharp rise but may not have a fixed duration, though its amplitude may be critical.

trimmer capacitor a small-value preset VARIABLE CAPACITOR or RESISTOR used for the fine adjustment of a CIRCUIT. A trimmer is adjusted only during manufacture or calibration of a circuit. One example of the use of a trimmer is in the oscillator TRACKING of a SUPERHETERODYNE receiver. A trimmer capacitor is connected in parallel with the main tuning capacitor, and another form of trimmer, known as a padder, is sometimes connected in series. By adjusting these capacitors, the tracking of the oscillator frequency can be made almost perfect at the two extremes of the frequency range.

Trinitron *Trademark.* the Sony aperture grille color TV tube. See SHADOW MASK.

triode see VACUUM TUBE.

trip a RELAY connected so as to open-circuit a supply in the event of a fault. See also OVERCURRENT TRIP.

triple-detection receiver see DOUBLE SUPERHETERODYNE.

tripler a form of AC VOLTAGE MULTIPLIER circuit whose output voltage is three times the input voltage.

Tri-state device or **Tri-state output** *Trademark.* a digital device that combines the speed of PUSH-PULL outputs with the interconnection ability of OPEN-COLLECTOR DEVICES by providing a control input that effectively disconnects an output when it is not required. See also THREE-STATE LOGIC.

TR switch see TRANSMIT-RECEIVE SWITCH.

true RMS the measurement of the true ROOT-MEAN-SQUARE value of an ALTERNATING CURRENT quantity. Many AC meters operate by rectifying the AC and measuring the resulting unidirectional current. The meter resistors can be adjusted so that for a sine-wave input, the readings will correspond to the values of RMS current or voltage. This is not a true RMS reading, however, because using the meter on any waveform other than a sinusoid, with a different FORM FACTOR, will give incorrect results. Only meters that operate on a square-law principle, such as a MOVING-IRON METER or a HOT-WIRE METER, will give true RMS values for any waveform.

trunk a main set of CABLES connecting from one unit to another.

truth table a table that shows all possible inputs and outputs for a LOGIC CIRCUIT.

T-section one half-section of a filter, the other being the pi-section. See Fig. 95.

TTL see TRANSISTOR-TRANSISTOR LOGIC.

T-type flip-flop a FLIP-FLOP circuit of the toggled (see TOGGLING) type. The T-type flip-flop has one input, the toggle input. Each pulse at this input will cause the output to reverse the output state between logic 0 and logic 1.

tube see CATHODE-RAY TUBE, VACUUM TUBE.

tuned antenna an antenna, such as a DIPOLE ANTENNA that is tuned in the sense of being cut to some multiple of the wavelength. This enables formation of a STANDING WAVE on the antenna.

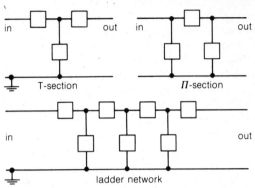

FIG. 95. **T-section.** The T- and pi sections of a filter, and the ladder network that results from joining the sections. The elements of the network are shown as impedances.

tuned amplifier an AMPLIFIER in which load RESISTORS are replaced by RESONANT CIRCUITS so that amplification is a maximum at the frequency of resonance of the tuned circuit in each stage. See also STAGGER TUNING.

tuned circuit see RESONANT CIRCUIT.

tuned filter a FILTER designed to accept or reject a very narrow frequency band. See also TRAP.

tuned-load oscillator a form of OSCILLATOR in which both INPUT (1) and OUTPUT of an AMPLIFIER are tuned to the same FREQUENCY. The FEEDBACK through STRAY CAPACITANCE ensures that the circuit will oscillate at that frequency.

tuner the portion of a RADIO RECEIVER comprising the RF, IF, and DEMODULATOR stages. A separate tuner is often part of a hi-fi system because incorporating a tuner and an amplifier in the same casing can generally lead to NOISE problems.

tuning indicator a DEVICE used to indicate the accuracy of manual tuning. At one time, a form of vacuum device known as the *magic eye* was used for this purpose, but in the few examples where tuning indicators are still used they consist now of meters, solid-state displays, or modern forms of VACUUM-FLUORESCENT DISPLAY.

tunnel diode a heavily doped P-N JUNCTION diode that contains a negative-resistance region in its current-voltage CHARACTERIS-TIC. Because of this negative resistance region, the tunnel diode will oscillate when biased into the appropriate region, and by connecting a suitable RESONANT CIRCUIT usually a resonant cavity, the oscillations can be tuned to any desired frequency. The tunnel diode is normally used as a very low power microwave or radar oscillator.

turn-off time the time for current to stop flowing in a THYRISTOR once the turn-off conditions have been reached.

turns ratio the ratio of secondary to primary turns in a TRANS-FORMER.

turret tuner a tuning system once used for TV receivers and FM tuners. It consisted of a cylinder that carried the tuning inductors inside with a set of connecting contacts outside. By rotating this cylinder inside a casing, the connections to each coil could be made, thus tuning from one TV channel to another.

tweeter a LOUDSPEAKER intended to reproduce high-frequency notes. Compare WOOFER.

twin lead or **twin line** a lead made from two insulated closely spaced conductors. The lead has low radiation loss when used as a TRANSMISSION LINE. Compare COAXIAL CABLE.

twin line see TWIN LEAD.

twin-T network or **parallel-T network** a type of FILTER network using resistors and capacitors that is frequency selective. The device is used as a FILTER or as the frequency-selecting part of an OSCILLATOR circuit.

twisted pair a cable consisting of a pair of conductors twisted around each other and carrying a signal, usually digital, and its complement. This results in increased noise immunity, and such cables are used when serial digital data have to be transmitted by cable over distances greater than a few meters.

twisted-ring counter a COUNTER circuit that uses J-K FLIP-FLOPS in a ring closed with one J-to-K and K-to-J connection.

two-phase (of an AC supply) using one live and one neutral line, as in normal house current.

TWT see TRAVELING-WAVE TUBE.

U

UHF *abbreviation for* ultrahigh frequency, the range of about 300 MHz to 3 GHz, 1 m to 10 cm wavelength. This range is widely used for television broadcasting.

ULA see UNCOMMITTED LOGIC ARRAY.

ultrahigh definition a development of television intended to enhance picture quality. In place of the existing line standards of 525 (USA and Japan) or 625 (Europe) lines per picture, ultrahigh-definition systems plan to use 1,000 or more lines per picture, abandoning existing standards of bandwidth and frequency.

ultrasonic (of an acoustic wave) having a frequency above the audible range. The ultrasonic range is usually taken as 20 kHz and above, but most ultrasonic devices use frequencies in the range 100 kHz to 5 MHz. The waves are generated by PIEZOELECTRIC CRYSTALS or, for the lower frequencies, magnetostrictive transducers (see MAGNETOSTRICTION).

ultrasonic delay line a DELAY device for signals that makes use of ultrasonic waves. Because ultrasonic waves travel through solids or liquids at speeds much lower than the speed of radio waves in space or along conductors, a short length of ultrasonic wave path can cause a delay of several microseconds. The signals are converted to ultrasonic form by a TRANSDUCER then travel through a solid or liquid, and are then converted back into electromagnetic waves by a second transducer.

Early ultrasonic delay lines used mercury as the wave medium. Later varieties have used glass and other solid materials. The main requirement for delay lines originally was in integration of signals for radar purposes, but a huge new market opened up when the PAL color TV system was adopted. PAL delay lines use glass, with a V-shaped signal path that permits delay time to be adjusted by grinding the reflecting surface at the apex of the V.

ultraviolet the range of electromagnetic wave frequencies lying between the violet end of the visible spectrum and the start of the x-ray region. Ultraviolet radiation is of higher energy than light and can have destructive effects on the eye and skin. One important application is in erasing PROMS.

unbalanced see SINGLE-ENDED.

unclocked (of a gate circuit or flip-flop) not controlled by a CLOCK pulse. See also ASYNCHRONOUS.

uncommitted logic array (ULA) an INTEGRATED CIRCUIT consisting of a large number of gates whose connections are not made at the time of initial manufacture. The connection pattern can be determined by the customer, using a large chart, and then applied to the integrated circuits by one final process. This gives flexibility of custom-designed ICs at almost the price of mass-produced chips.

undercoupling see LOOSE COUPLING.

undercurrent release a CONTACT BREAKER that operates when the current in a circuit is below a specified level. See also TRIP, OVERCURRENT TRIP.

underdamped (of a RESONANT CIRCUIT) oscillating for a few cycles after a sudden voltage change. Compare DAMPED, OVER-DAMPED.

underscan a SCAN of a TV CAMERA TUBE or receiver tube with less than normal amplitude. This is usually avoided because of the danger of creating a visible boundary at the edge of the scan.

undershoot a transient decrease of voltage at the trailing edge of a PULSE. Compare OVERSHOOT.

undervoltage release a CONTACT BREAKER that operates when the voltage in a circuit is below a set level. See also TRIP, UNDER-CURRENT RELEASE.

undistorted wave a WAVE whose WAVEFORM has not been changed in an electronic circuit.

unidirectional (of a CURRENT usually rectified alternating current) traveling in one direction.

uniform line a TRANSMISSION LINE whose constants of CAPACI-TANCE, INDUCTANCE, and CHARACTERISTIC IMPEDANCE are the same throughout its length.

uniform waveguide a WAVEGUIDE whose CHARACTERISTICS are the same throughout its length.

unijunction or **double-base diode** a type of pulse-generating semiconductor device. The unijunction consists of a bar of lightly doped N-TYPE silicon with an OHMIC CONTACT at each end. Between these ohmic contacts, a junction is formed to the emitter, a heavily doped p-region. Resistance between the two ohmic contacts is very high until carriers are injected from the junction.

The unijunction, therefore, is nonconducting until the voltage at the emitter reaches a certain fraction of the voltage between the ohmic connections. This fraction is called the INTRINSIC STAND-OFF RATIO. When the emitter voltage reaches the value that causes conductivity, comparatively large current can flow both between the emitter and the grounded base contact, and between the two base contacts. The main application is in pulse generators and sawtooth oscillators. See Fig. 96.

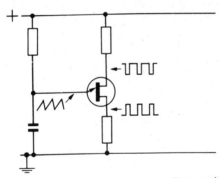

FIG. 96. **Unijunction.** A simple unijunction oscillator and the waveforms at the electrodes.

unilateral network a NETWORK that conducts in one direction only. This normally applies to signals, so that a transistor is a unilateral network in the sense that the direction of signal is from input to output, not the reverse. This is true only if the collector-base capacitance is low and the collector-base resistance is high. At very high frequencies, a transistor may allow signal transfer through the collector-base capacitance, and the correction of this unwanted feedback is called *unilateralization*. See also NEUTRALIZATION.

unipolar (of a SEMICONDUCTOR device) having the main current carried by MAJORITY CARRIERS in one polarity of material, without passing through a JUNCTION. The FIELD-EFFECT TRANSISTOR is one example of a unipolar device.

unit a physical quantity of strictly defined and reproducible size. The fundamental units are the kilogram, meter, second, and ampere. The lumen is used for light measurement.

unit-step a sudden change of VOLTAGE between specified limits, such as logic 0 level to logic 1 level.

universal motor an electric motor that operates on either AC or DC. A universal motor uses an electromagnetic field coil rather than a permanent magnet, and a commutator to feed the rotor coils. Compare AC MOTOR.

universal shunt a meter SHUNT that is tapped so that the ratio of series to shunt resistance in the current path can be varied. This enables a MOVING-COIL METER connected to the shunt to be used for a whole set of current ranges.

unloaded Q the value of Q FACTOR for a RESONANT CIRCUIT that is not supplying any form of load.

untuned antenna an ANTENNA whose dimensions are unrelated to the wavelength of the signal, as, for example, a long-wave transmitting or receiving antenna.

untuned circuit see APERIODIC CIRCUIT.

up-down counter a logic COUNTER based on FLIP-FLOPS connected by GATES (2). The gates are controlled by a single line so that one voltage level on the line produces up-counting (incrementing), and the opposite level produces down-counting (decrementing).

upper sideband the SIDEBAND of a modulated carrier whose frequencies are higher than the carrier frequency.

V
---

vacuum capacitor a CAPACITOR that uses a vacuum as its insulating dielectric. Vacuum capacitors are widely used in transmitting circuits, because they can be made variable by varying their geometry and because they do not suffer from nonlinear effects.

vacuum evaporation a method of depositing thin films, mainly of metals. The material to be deposited is attached to a tungsten filament that can be heated electrically. The surface to be coated is arranged at about 10—15 cm distance, and the assembly is enclosed and evacuated. When the filament is heated, the material on it evaporates and condenses as a thin film on the cool surface. See also METALIZING, SPUTTERING.

vacuum fluorescent display any type of display for ALPHANU-MERIC characters that uses a CATHODE emitting electrons that strike shaped anodes coated with a phosphorescent material. Vacuum fluorescent displays have been widely used for pocket calculators and for metering indicators. They enable reasonably high brightness to be achieved at low power levels and are often more satisfactory in this respect than LEDs. Their disadvantages are relative fragility and the need for a supply voltage higher than normally used in digital equipment.

vacuum switch a SWITCH whose CONTACTS are in a vacuum so as to avoid contact oxidation through sparking.

vacuum tube an evacuated device that contains a thermionic cathode and other electrodes. The most common form of vacuum tube is the triode transmitter tube, which is the thermionic equivalent of a BIPOLAR TRANSISTOR. It uses a directly heated FILA-MENT surrounded by a framework carrying a metal spiral, the control grid. In a *triode* this in turn is surrounded by a metal cylinder, the ANODE, or plate. For high-power tubes, the anode is the exterior casing of the tube and is finned on the outside, or attached to a water jacket for cooling. Small tubes contain the anode and other electrodes inside a glass ENVELOPE.

A vacuum tube that uses one grid is a triode, and adding another grid between the control grid and the anode makes the tube a *tetrode*. The cathode is heated to provide electrons, and the anode is connected to a high voltage, typically at least 1 kV. The flow of electrons is controlled by the control grid voltage, and a sufficiently negative voltage on this electrode will cut off the electron stream. For transmitting purposes, the grid voltage is often swung between negative and positive voltage values relative to the cathode, and current will flow to the grid only during the positive parts of the cycle.

vacuum tube see THERMIONIC VALVE.

valley point the minimum-current point in a TUNNEL DIODE characteristic. See Fig. 97.

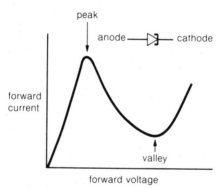

FIG. 97. **Valley point.** The valley point of a tunnel diode. The region of negative resistance is between the peak and the valley.

vapor cooling a method of cooling transmitter tubes. The ANODE is surrounded by a jacket through which vapor, usually water vapor, is circulated. The vapor can be circulated much faster than water and provide a more efficient method of transferring heat.

varactor a SEMICONDUCTOR diode with voltage-variable CAPACITANCE. The diode is operated with REVERSE BIAS and the effect of the bias on the DEPLETION LAYER at the junction makes this layer act like a variable thickness of dielectric. The capacitance is therefore smallest with large reverse bias, and largest with small reverse bias.

varactor tuning a type of tuning employing an INDUCTOR and a VARACTOR. The varactor is normally connected in series with another capacitor to ensure DC isolation from the tuned circuit. Varactor tuning is widely used at VHF and UHF, where only small variations of capacitance are needed, and where AUTOMATIC FREQUENCY CONTROL circuits are essential.

variable capacitor a CAPACITOR whose value is mechanically variable. The usual variation method is to have a set of (fixed) STATOR plates and a set of (rotating) ROTOR plates whose meshing can be altered by rotating the shaft to which the rotor plates are connected. The rotor plates normally are grounded to avoid problems of BODY CAPACITANCE. TRIMMER CAPACITORS can use compression, in which a threaded bolt tightens or loosens plates that clamp a solid dielectric. The *beehive trimmer* uses concentric rings for rotor and stator and varies the interaction by screwing the rotor plates into or out of the stator plates. Variable capacitors can be ganged (see GANGED CIRCUITS).

variable inductor an INDUCTOR whose value can be varied mechanically. The method of variation is normally to alter the position of a high-permeability core relative to the inductor windings.

variable reluctance the principle used for the MOVING-IRON PICKUP. The movement of an armature that is part of a magnetic circuit alters the reluctance, which is the magnetic-flux equivalent of resistance.

variable resistor a RESISTOR whose RESISTANCE value can be changed mechanically. The construction is as for a POTENTIOMETER but only one end and the moving tap of the potentiometer are used.

Variac *Trademark.* a variable AUTOTRANSFORMER.

variometer a seldom used INDUCTOR whose value of SELF-INDUCTANCE is variable.

varistor a nonohmic resistor (see OHMIC). Varistors are made from semiconductor material, have an exponential current-voltage characteristic, and are used as overvoltage limiters, because the current through the varistor will greatly increase for a very small increase in the voltage across it.

V band the MICROWAVE band of wavelength range 53.6 to 63.2 mm, frequency range 46 to 56 GHz.

VDU see VISUAL DISPLAY UNIT.

velocity microphone a MICROPHONE operated by the velocity of air moving past it rather than by pressure. The essential feature of a velocity microphone is that the sound waves can strike all sides of the vibrating element. See also RIBBON MICROPHONE.

velocity modulation the interaction between an ELECTRON BEAM and an electromagnetic wave. The velocity of electrons in the beam is modulated by the wave, causing the beam of electrons to become bunched. See KLYSTRON, TRAVELING-WAVE TUBE, CARCINOTRON, BACKWARD-WAVE OSCILLATOR.

Veroboard *Trademark.* a BREADBOARD system. The original Veroboard is a PRINTED CIRCUIT BOARD consisting of parallel strips of copper on an insulating sheet. The strips are drilled at regular intervals (originally 0.15″, later 0.1″) so that components can be inserted and soldered to the board. Strips can be cut across where necessary for isolation. Veroboard enables experimental circuits to be made and tested in a form that is closer to the final production shape than is usually possible with breadboard systems.

vertical blanking the suppression of video signal in the time of the field or frame flyback. See RASTER, VERTICAL TIMEBASE.

vertical hold a TELEVISION RECEIVER control for synchronizing the VERTICAL TIMEBASE.

vertical timebase the TIMEBASE circuit in a TELEVISION RECEIVER that generates the field (vertical scan) waveform. The WAVEFORM consists of a current sweep with a duration of about 20 milliseconds, followed by a rapid flyback.

vestigial sideband a method of TV signal transmission. The use of single-sideband SUPPRESSED-CARRIER TRANSMISSION methods, though efficient, is impossible for TV signals because of the severe distortion that would be produced in the video signals. A compromise is to attenuate the modulated signal in such a way that the carrier amplitude is only half its original amplitude, and the lower sideband contains only the lower frequencies of the video signal.

This makes the lower sideband vestigial, and enables more efficient use of the transmitter power. See Fig. 98.

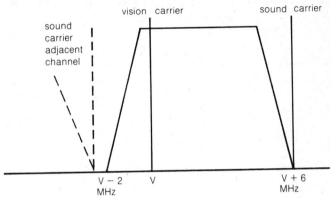

FIG. 98. **Vestigial sideband.** The upper sideband is allowed to extend for its full 6 MHz extent, but the lower sideband is filtered so as to extend only to 2 MHz below the vision carrier frequency. This represents a good compromise between bandwidth conservation and distortion problems.

VHF *abbreviation for* very high frequency, the range of approximately 30 to 300 MHz, 10 to 1 m wavelength.

vibrator see CHOPPER.

video relating to or employed in the transmission of picture information, that is, TV camera signals or any other signals that can produce an image. The VIDEO SIGNAL is of wide bandwidth and may be a COMPOSITE SIGNAL with SYNCHRONIZING PULSES included.

video amplifier a type of BROADBAND amplifier for video signals. For TV purposes, the video amplifier need provide only for unidirectional signals, so TV video amplifiers can use AC coupling with DC restoration at the output. For oscilloscopes and other instruments, video amplifiers may have to be directly coupled, capable of amplifying pulses of either polarity, and with an upper frequency limit of 30 MHz or more.

video cassette recorder a system that uses a cassette containing MAGNETIC TAPE for TV signal recording. The two competing commercial systems for domestic use are Sony's Betamax and JVC's VHS, though the recent 8 mm standard may supplant both. Half-inch wide tape is used, and the tape speed is in the region of 1.8 cm/second for Beta, 2.34 cm/second for VHS.

The recorder can take signals directly from an anetnna and has normal receiver circuits that will produce the LUMINANCE (black/white), chroma (color), and sound signals. The luminance signal is frequency-modulated onto a carrier of about 6 MHz. The color signals are frequency-shifted to a lower frequency range and then added to the FM luminance signals. The use of FM enables the luminance signals to be recorded at a constant amplitude, thus avoiding the problems of the limited amplitude range of tape. The record/replay heads rotate, and the tape direction is skewed so that the recording tracks are across the tape at a shallow angle, permitting a high rate of tape scanning for a slow rate of tape movement.

The BANDWIDTH of the system is limited, so the resolution of the replayed picture is lower than that of a transmitted picture, typically about half the resolution in terms of lines. The audio signal is separately recorded on a track that lies parallel to the length of the tape (see RECORDING OF VIDEO). A synchronizing control track is recorded in the same way on the other edge of the tape. On replay, the two sets of video signals have to be reconstituted and modulated onto another UHF carrier, and the sound signal is also modulated so that the complete UHF video output can be used by any home receiver. Most recorders feature a composite video output so a monitor can be used for a quality display. The more recent 8mm standard offers better quality of video, greatly improved sound quality, and compatibility with miniature video cameras.

See also DYNAMIC TRACKING.

video frequency 1. a frequency in the video range of about DC to 5 MHz. **2.** *adj.* (of a circuit) being designed to cope with video signals.

video mapping a computer technique applied to a RADAR display. A map or photographic image of an area is superimposed over a radar trace so that the geographical position of targets can be seen.

video recording see RECORDING OF VIDEO.

video signal a SIGNAL that can be used to provide a picture. The term usually denotes a COMPOSITE SIGNAL containing the video waveform along with color signals and all the synchronizing signals.

videotape a form of MAGNETIC TAPE for recording VIDEO SIGNALS. The tape is wider than audio tape, typically half-inch for video cassettes, and 50 mm, just under 2 inches, for professional machines for broadcast use. The coating material is usually ferric oxide with cobalt oxide.

vidicon the almost universally used TELEVISION CAMERA tube. A vidicon makes use of a photoconductive semiconductor, such as antimony trisulfide or lead oxide. The optical image from the camera lens is projected onto the flat faceplate of the vidicon, typically one inch in diameter, although half-inch wide vidicons are used in miniature cameras. Inside the faceplate, a transparent conducting layer on the glass forms the output terminal, with a connection at the edge of the faceplate. The conducting layer is coated with the photoconducting material, the TARGET which in turn can be scanned by a low-velocity ELECTRON BEAM.

An optical image causes a corresponding pattern of low and high conductivity paths in the photoconductor, and these allow the inner surface to charge from the steady voltage at the target. The scanning electron beam discharges these parts of the inner surface, causing a signal voltage to appear by capacitive coupling on the conducting side. This signal voltage is the video signal out of the vidicon. See Fig. 99.

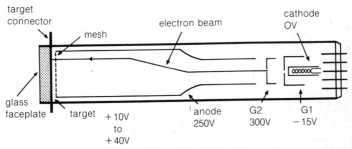

FIG. 99. **Vidicon.** See this entry.

virtual cathode an area of SPACE CHARGE that behaves like a supply of electrons.

virtual ground 1. a point in a CIRCUIT at which there is no SIGNAL at any time, though the point is not directly connected to ground. 2. the INPUT (1) to an amplifier in which large amounts of negative feedback are used.

visual display unit (VDU) a CATHODE-RAY TUBE display or large screen LCD display for a computer.

VLF *abbreviation for* very low frequency, the range of approximately 3 to 30 kHz, 100 to 10 km wavelength. These frequencies are not particularly useful for surface radio use, but allow some degree of underwater communication.

VLSI *abbreviation for* very large scale integration, that is, many thousands of active devices per chip. The search is on for a suitable phrase to express even larger scales of integration. See also LSI, MSI.

voice coil the active part of a LOUDSPEAKER. The voice coil is located at the center of the speaker cone, where it is supported in the field of a magnet by the cone of the loudspeaker. An alternating current through the voice coil causes movement to and fro, which displaces air, thus forming a sound wave.

volatile (of the MEMORY of a SEMICONDUCTOR) not retaining stored information when the power supply is switched off. Compare NONVOLATILE. See also ROM, MAGNETIC MEMORY.

volt the unit of ELECTRIC POTENTIAL and POTENTIAL DIFFERENCE. The volt is defined as the joule per coulomb, the amount of work done per unit charge in moving a charge from one place to another.

voltage see ELECTRIC POTENTIAL.

voltage amplifier an AMPLIFIER that has VOLTAGE GAIN. The output signal from a voltage amplifier will be a copy of the input signal, but with greater voltage amplitude.

voltage doubler a RECTIFIER circuit that provides double the peak voltage of the AC input. This is twice as much as would be obtained from a conventional half- or full-wave rectifier. See also VOLTAGE MULTIPLIER.

voltage drop the VOLTAGE between the ends of a CONDUCTOR or a COMPONENT. The voltage drop is due to the RESISTANCE or IMPEDANCE of the conductor or component when a current flows.

voltage feedback the FEEDBACK of a SIGNAL that is a selected fraction of the OUTPUT voltage signal to the input of a circuit.

voltage gain the ratio of OUTPUT signal voltage to INPUT (1) signal voltage for an AMPLIFIER. The voltages must be measured in the same way, both peak-to-peak or both RMS, and can be converted into decibels. Voltage GAIN takes no account of the phase between output and input.

voltage level the signal VOLTAGE at some point in a CIRCUIT compared with a standard level. For VIDEO SIGNALS for example, the standard level is one volt peak-to-peak, ignoring synchronizing pulses.

voltage multiplier a RECTIFIER circuit whose DC output may be several times the peak voltage of the AC input. The technique uses a chain of DIODES and CAPACITORS with each rectifier having as its input the DC level of the previous rectifier, plus the original AC signal attenuated only slightly by the capacitor chain.

voltage reference diode a diode using AVALANCHE BREAK-DOWN. The diode is usually called a ZENER DIODE and is catalogued as such. The voltage reference diode is operated with RE-VERSE BIAS and the voltage across the diode is almost independent of the current. By using the diode with a current that is already partly stabilized, a precisely constant voltage value can be generated. The reference diode is used in VOLTAGE STABILIZERS.

voltage regulator see VOLTAGE STABILIZER.

voltage stabilizer or **voltage regulator** a circuit whose output is a steady voltage. Within limits, the output voltage should not be affected by differing load currents or by changes in the AC supply voltage. Most voltage stabilizers make use of NEGATIVE FEEDBACK amplifiers with direct coupling. A fraction of the output voltage is compared with the voltage of a VOLTAGE REFERENCE DIODE and any difference is used to correct the output. The simplest designs make use of an EMITTER FOLLOWER with the reference diode in the base circuit. Voltage stabilizers can be of series or shunt design, but are overwhelmingly of the series type.

voltage standing-wave ratio (VSWR) the STANDING-WAVE RATIO measured from signal voltage amplitude levels.

volt-ampere the product of VOLTS and AMPERES for a SIGNAL. Even if both quantities are measured as RMS quantities, the product is not necessarily equal to dissipated power because of PHASE DIFFERENCES. To obtain power, the volt-ampere figure must be multiplied by the POWER FACTOR which is the cosine of the phase angle.

voltmeter a METER for measuring VOLTAGE. The predominant type is the MOVING-COIL METER with a set of DC ranges. Certain AC measurements are made possible by using a RECTIFIER but these readings are of lower precision and are affected by the frequency and waveform of the AC. Increasingly, DIGITAL voltmeters are replacing the moving-coil type for DC measurements, and OSCILLOSCOPES are more satisfactory for work with signals. Voltmeters with MOSFET inputs (see FIELD-EFFECT TRANSISTORS) can be used for making measurements that require exceptionally high input resistance. These meters have replaced the old vacuum tube voltmeters and the even older electrostatic voltmeters.

volume 1. the AMPLITUDE of sound as heard by the ear. For a sound of constant amplitude level, the effect on the ear is not constant but varies with the frequency because of the NONLINEAR response of the ear. **2.** the amplitude of an AUDIO SIGNAL.

volume compressor a CIRCUIT that reduces the range of amplitudes in an AUDIO SIGNAL. This is done typically by a nonlinear amplifier, whose gain is inversely proportional to the signal amplitude. Volume compression is extensively used in sound recording, because no recording system or broadcasting system can cope with the full amplitude range (dynamic range) of orchestral music. See also WHITE COMPRESSION, VOLUME EXPANDER, COMPANDER, DOLBY SYSTEM, DBX.

volume control the POTENTIOMETER control on sound reproduction equipment for adjusting the intensity of sound by altering the amplitude of audio signals.

volume expander a CIRCUIT that increases the AMPLITUDE range of a SIGNAL. The volume expander is a NONLINEAR amplifier that has higher gain for larger amplitude signals. Volume expanders are usually designed in conjunction with VOLUME COMPRESSORS to form COMPANDERS so that a signal that has been compressed and then expanded will be restored to normal amplitude range. Companding is extensively used in tape recording and optical film recording systems. See DOLBY SYSTEM, DBX.

volume limiter a CIRCUIT that prevents the AMPLITUDE of a SIGNAL from exceeding a preset limit. This is commonly used for TELEPHONY so that overmodulation of the carrier, which causes severe sideband distortion, is impossible. Crude limiters clip the waveform, but modern types use a NONLINEAR amplifier that affects only the amplitude range near the limit.

von Neumann architecture the conventional method used in computer design, in which each instruction is carried out in sequence.

The alternative is some form of parallel architecture. Research is being carried out to solve the problems of parallel architecture, in which several instructions can be carried out simultaneously, and to devise the computing languages that will be required to make efficient use of parallel architecture. Named for John von Neumann (1903—57), US mathematician.

VSWR see STANDING WAVE RATIO.

VU meter volume-unit METER a form of recording-level meter for a tape recorder. The meter measures the average signal level and should give an indication that is proportional to the power level of the signal, thus allowing for the differences between smooth and spiky waveforms.

W

---

wafer a thin circular slice of a large crystal of SEMICONDUCTOR material. Wafers of 3″ and 4″ diameters are processed by PHOTOLITHOGRAPHY so that several hundred CHIPS can be formed in one set of operations. After the wafer processing, the individual chips are tested automatically by a machine that marks the rejects, and the chips are separated by ruling the wafer with a diamond cutter. The good chips are then selected and passed on to headers for mounting.

walkie-talkie a miniature portable sound transmitter-receiver.

warble a small variation in FREQUENCY usually at an AUDIO rate. This can be due to speed variations in the motor of a film projector, record player, or tape recorder.

warmup the time after switch-on during which the behavior of a CIRCUIT changes. Warmup time for a circuit that contains a CATHODE-RAY TUBE includes the time needed to heat the cathode until its electron emission is steady.

watt the unit of POWER. Electrically, the power value in watts is obtained by multiplying the DC voltage value in volts by the DC current value in amperes. For AC or signal quantities, ROOT-MEAN-SQUARE values must be used, and any differences in phase expressed by a POWER FACTOR correction.

wattless component see REACTIVE COMPONENT.

wattmeter an instrument that measures electrical power.

wave an alternating quantity whose effect can spread through a material or through space. For electronics purposes, the most important waves are electromagnetic waves, which are the familiar radio waves used as CARRIERS. See also ULTRASONIC.

wave analyzer an instrument, based on a CATHODE-RAY OSCILLO-SCOPE display, that analyzes a wave into its components. A typical output from a wave analyzer is a set of vertical bars in which the height of each bar indicates the amplitude of a sine wave, and the bar position indicates frequency. In this way, the wave analyzer shows the amplitudes of the fundamental and each HARMONIC for a complex waveform. Cathode-ray displays now are being replaced by computer displays. See also FOURIER ANALYSIS.

waveform or **waveshape** a graph plot of wave AMPLITUDE against time for the time of one CYCLE. The common waveforms are sine, square, pulse, and triangle.

wavefront the line of advance of a WAVE through a material. The wavefront of a wave from a point source, for example, is in the shape of a sphere. The wavefront consists of all the points at which the wave has traveled through the material for the same time.

waveguide a hollow CONDUCTOR for MICROWAVES. Because of radiation and surface resistance, microwaves cannot be conducted efficiently along conventional open lines. A waveguide, consisting normally of a metal tube of rectangular cross section, is much more efficient. The metal can be silver-plated to increase surface conductivity, and the wave is held inside the guide by reflections from the wall. The dimensions of the waveguide have to be chosen appropriately for the wavelengths to be propagated, and the ratio of width to depth for the cross section will determine the pattern (see MODE) of electric and magnetic fields inside the waveguide.

wave heating the heating of materials by means of MICROWAVE bombardment. This has been used for rapid drying of damp materials, setting adhesives, and cooking food. See also MICROWAVE OVEN.

wavelength the distance between one wave peak and the next in a wave. The wavelength is a particularly easy feature of a wave to measure (see LECHER WIRES), and its size determines many of the features of the wave, such as ease of PROPAGATION. The value of wavelength multiplied by FREQUENCY gives speed of propagation in the medium in which wavelength is measured.

wavemeter a form of frequency METER. A RESONANT CIRCUIT is adjusted so as to resonate with an incoming wave, and the resonant current is displayed on a meter. The frequency can then be read from a dial. The wavemeter can be calibrated in terms of wavelength, assuming that the waves are in the air.

waveshape see WAVEFORM.

wave train a set of WAVES usually implying a limited number.

wave trap a form of wave FILTER that rejects a particular frequency.

W band the range of 3 to 5.36 mm wavelength, 56 to 100 GHz microwave frequency.

Wheatstone bridge the original and classical form of BRIDGE circuit used for resistance comparison.

whistle an INTERFERENCE effect caused by another CARRIER. The beating of two carriers close to each other in frequency causes a whistle to be heard at the receiver. The *whistle frequency* is the difference between the carrier frequencies.

white compression a form of volume compression (see VOLUME COMPRESSOR) on a VIDEO SIGNAL. It compresses the amplitudes that correspond to the white (brightest) areas of the picture.

white noise an electrical or other form of NOISE that has a broad spectrum of FREQUENCIES and a uniform AMPLITUDE over the entire frequency range.

white peak an AMPLITUDE peak in a VIDEO SIGNAL that corresponds to a bright white part of the picture.

wideband see BROADBAND.

Wien bridge a form of BRIDGE circuit used mainly as a filter or as a tuning circuit for a low-frequency oscillator. The circuit was originally used as a measuring bridge for frequency. See Fig. 100.

Winchester disk or **Winchester** a permanently mounted type of HARD DISK used for small computer systems.

winding a length of coiled CONDUCTOR used to produce a MAGNETIC FIELD or have a VOLTAGE induced in it by a magnetic field. See also PRIMARY WINDING, SECONDARY WINDING.

window 1. a material transparent to radiation. 2. the range of FREQUENCIES that can pass through the IONOSPHERE from outer space, the radio window. 3. the thin mica cover of a GEIGER-MÜLLER TUBE, which allows ionizing particles and radiation to pass into the tube. 4. the faceplate of a TV camera tube. 5. a vacuum-sealed connection between a MICROWAVE GENERATOR and a WAVEGUIDE.

wire 1. a metal CONDUCTOR in wire form. 2. any connecting conductor that is not a strip in a PRINTED CIRCUIT BOARD.

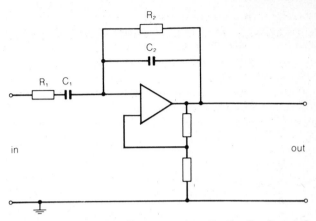

FIG. 100. **Wien bridge.** The components R_1, R_2, C_1, C_2 are the Wien bridge components.

wire broadcasting the distribution of SIGNALS at radio frequencies by wire rather than through space.

wired-OR a method of connecting OUTPUTS from integrated circuit gates. Some types of gates with open-collector outputs (see OPEN-COLLECTOR DEVICE) can have their outputs connected to a common load. This allows the output voltage to be pulled down to logic 0 if any one of the gates conducts, giving an OR logic action. Gates of conventional PUSH-PULL transistor-transistor logic stages may be destroyed if they are connected together in this way.

wireless an old term for RADIO shortened from wireless (meaning without CABLES) telegraphy.

wire-wound resistor a type of RESISTOR constructed by winding fine-gage insulated wire around an insulating former. The power dissipation of such a resistor can be large, but it is difficult to make the resistor noninductive.

wire wrapping a prototype construction method in which a special kind of wire is wrapped tightly around a solder-coated pin. When using this technique, no heat is applied to the pins. Connections are made solely by the tight contact achieved by the wrapped wire.

wiring diagram a diagram showing the physical connection of wires in a circuit. Compare CIRCUIT DIAGRAM.

woofer a LOUDSPEAKER particularly suitable for reproducing BASS notes. A woofer uses a large, freely suspended cone. At high frequencies, such a cone would resonate, with standing waves forming on it, so the woofer is isolated from high frequencies by means of the CROSSOVER NETWORK. See also TWEETER, LOUDSPEAKER.

work function the amount of energy that is needed to liberate one ELECTRON from a material, usually quoted in terms of ELECTRON-VOLTS.

wound core a TRANSFORMER core consisting of a ribbon of magnetic material wound into a core shape.

wow a slow frequency modulation of sound from a disk or tape source. Wow is caused by uneven disk or tape speed. Compare FLUTTER.

write store data in a MEMORY. Compare READ.

X Y Z

x-axis the horizontal axis of a graph or of a CATHODE-RAY TUBE display.

x band the MICROWAVE band range of 27.5 to 57.7 mm, 5.2 to 10.9 GHz.

xerography a method of copying text or graphics that relies on ELECTROSTATIC charging and PHOTOCONDUCTIVITY. The principle is also used in laser printers for computers.

XOR gate see EXCLUSIVE OR.

x-plates the horizontal DEFLECTION PLATES of an INSTRUMENT TUBE. The timebase waveforms are normally applied to these plates.

x-ray the range of electromagnetic radiation with WAVELENGTHS much shorter than those of visible light. X-rays penetrate most materials and cannot be focused, reflected, or refracted by conventional methods.

x-ray tube a vacuum tube in which X-RAYS are generated. The tube generates a very high energy ELECTRON BEAM that is focused onto a metal anode, the TARGET. The bombardment of this target produces x-rays. The wavelength of the x-rays can be reduced by using higher voltages between cathode and target.

x-y plotter an instrument that produces a two-dimensional plot on paper. The plot is similar to that produced by an OSCILLOSCOPE but the x-y plotter is more useful if the rate of production of the pattern is slow.

Yagi antenna a type of DIRECTIONAL ANTENNA consisting of a DIPOLE ANTENNA fitted with a REFLECTOR and several DIRECTORS. The direction of the Yagi is the heading from reflector to director. The GAIN as compared with a simple dipole is large, and the directional properties of the antenna are useful in overcoming interference problems. The antenna is named for Hidetsugu Yagi (1886—1976), a Japanese engineer.

y-axis the vertical axis of **1.** a graph. **2.** the TRACE on a CATHODE-RAY TUBE.

yttrium-iron-garnet (YIG) a magnetic crystal of the FERRITE type. It is used in microwave circuits and for magnetic BUBBLE MEMORY applications.

yoke a set of DEFLECTION COILS for a TV CATHODE-RAY TUBE.

y-parameters a set of PARAMETERS formerly used for measuring TRANSISTOR characteristics.

y-plates the vertical DEFLECTION PLATES for an INSTRUMENT TUBE.

zener breakdown the reverse BREAKDOWN of a DIODE in which the p and n regions are heavily doped. Zener breakdown is more gradual than AVALANCHE BREAKDOWN but the name zener is used to denote any diode employing reverse breakdown.

zener diode any VOLTAGE STABILIZER diode that operates with reverse bias. The breakdown voltage of these diodes is almost independent of the amount of reverse current, so they can be used as voltage reference sources. See Fig. 101. See also VOLTAGE REFERENCE DIODE.

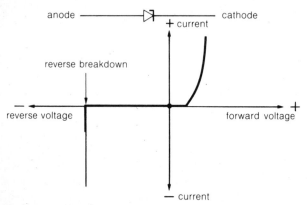

FIG. 101. **Zener diode.** A typical zener diode CHARACTERISTIC. The forward characteristic is normal, but the reverse characteristic exhibits an abrupt breakdown, so that the reverse voltage is almost constant over a wide current range after breakdown.

zero the minimum point on the complex S-plane of a POLE-ZERO DIAGRAM.

zero error a METER fault in which the needle of the meter does not return to zero. Instruments that use digital displays may also exhibit zero errors so that the reading is not zero with the meter disconnected or short-circuited.

zero frequency (zf) the frequency at which alternating current is absent, and direct current is steady.

zero level the level of comparison for SIGNALS. This may be zero voltage or the voltage of the signal baseline.

zero-level trigger a TRIGGER PULSE generated when an AC wave passes through the zero voltage level. Zero voltage triggers are used when a THYRISTOR is being switched in such a way as to generate little or no interference.

zero potential see GROUND POTENTIAL.

zf see ZERO FREQUENCY.

z-modulation the MODULATION of the ELECTRON BEAM of a CATHODE-RAY TUBE therefore modulating screen brightness. The term denotes particularly the modulation of an INSTRUMENT TUBE since TV cathode-ray tubes are normally modulated in this way.

z-parameters a set of PARAMETERS formerly used for measurement of TRANSISTOR characteristics.